New Guinea Gardens

New Guinea Gardens

A Study of Husbandry with Parallels in Prehistoric Europe

AXEL STEENSBERG

International Secretariat
National Museum, Brede, 2800 Copenhagen-Lyngby, Denmark

1980

ACADEMIC PRESS

A Subsidiary of Harcourt Brace Jovanovich, Publishers
London New York Toronto Sydney San Francisco

ACADEMIC PRESS INC. (LONDON) LTD
24–28 Oval Road,
London NW1

U.S. Edition published by
ACADEMIC PRESS INC.
111 Fifth Avenue,
New York, New York 10003

British Library Cataloguing in Publication Data

Steensberg, Axel
New Guinea gardens.
1. Food habits—Papua New Guinea
2. Man, Primitive—Food
3. Vernacular architecture—Papua New Guinea
I. Title
301.5′2 GN671.N5 80–49983

ISBN 0–12–664940–5

Filmset by Northumberland Press Ltd,
Gateshead, Tyne and Wear
Printed by Fletcher and Son Ltd, Norwich

Preface

This book is essentially the record of three short visits to Papua New Guinea in 1968, 1971 and 1975 made by someone who has spent his professional life in the archaeological, historical and ethnographic study of the material conditions of life and work amongst temperate European and particularly Danish peasants. It deals with aspects of the tools and technology associated with land clearance, gardening, fencing, housing and cooking in a small number of communities almost exclusively in the New Guinea Highlands. I have not attempted to present my observations in the context of an exhaustive review of the literature on these activities for New Guinea as a whole, let alone the broader geographical area of which New Guinea forms part. I have employed a different and somewhat unusual approach.

During my work in New Guinea, despite the vast and manifold differences of its inhabitants and their environment from anything I had experienced before, I continually felt myself in the presence of ways of going about the tasks of daily life that were familiar to me from my researches on the northern European peasantry. Common problems of getting food and providing shelter have evoked surprisingly similar responses in diverse areas of the world in pre-industrial conditions, where the family group was directly engaged in these concerns by its own efforts and from local resources. This book attempts to illustrate this proposition by setting my observations in New Guinea alongside historical and archaeological evidence from temperate Europe and allowing them to cross-fertilize. This is done in terms of the five themes of my New Guinea research set out in the first paragraph.

In a sense, therefore, the book is also an essay in what has recently come to be called ethno-archaeology, the controlled use of ethnographic observation to explain the archaeological record. I have attempted to show how much prehistorians can learn from technical observations of non-industrial life in areas like New Guinea. The handling of digging sticks and tools for cutting and hewing is a clue to the understanding of often slight differences in shape and size of such implements and to different methods by which they were hafted. The relationship of shape to function has always been a major concern of archaeologists and the ethnographic context is one in which it can be explored. For this purpose the camera is as necessary as the notebook, and at times the movie camera instead of the still camera. Thus while the hafting of a stone blade at Puyakira'go described in Chapter 1 could be adequately documented in 86 still photographs, its subsequent use in felling a tree could only be recorded on movie film. In the present publication the photographs are not a supplement to the text: they are an integral part of it.

Many anthropologists and not a few archaeologists have no experience of recording technology and technical operations and think themselves incapable of doing so. It is an aim of this book to show that this is a mistaken opinion. Everybody can cover agricultural work and household industries with a still camera and, where necessary, a simple movie camera. Back home there will always be trained people to help with the interpretation of the filmed record. Similarly the measurement of a house and the documentation of its construction and use, if done methodically, is neither an excessively technical nor time-consuming undertaking.

There are a number of reasons why records of these sorts should be carried out more frequently and by a wider range of scholars than they normally are. They can provide important leads to the understanding of aspects of culture other than the purely technical and material, namely the economic, social, spiritual and intellectual aspects. They will supply data of interest to students of technology and material culture which might otherwise remain uncollected. Finally, and very importantly, they will ensure knowledge of traditional procedures which are liable to change, even disappear, as societies are more and more integrated into or influenced by the modern industrial world which erodes the large degree of local self-sufficiency to which traditional ways of life were attuned. This account shows how changes have taken place in New Guinea Highlands technology in the relatively short time—45 years and less depending on the area—that Europeans have been there. The old people complain that the younger generation is growing up unaware of many of the old ways, and they are anxious that such traits should be recorded before they are lost. This is a wish shared by the leaders of the new Papua New Guinean state, and there are a number of institutions–the National Museum, the Institute of Papua New Guinea Studies and the National Arts School—charged with the record, preservation and development of different aspects of the varied cultural heritage of the country's many peoples. It is a huge task and can only be accomplished if scholars are prepared to step beyond the usual boundaries of their research, when in their privileged position in the field.

In 1968 my old student and colleague Jack Golson, now head of the Department of Prehistory, Research School of Pacific Studies of The Australian National University, invited me to visit New Guinea before I returned to Denmark from a congress I was to attend in Tokyo. He and other staff and students of his Department had recently begun archaeological work in New Guinea, and he was well aware that the traditional tools and techniques of agriculture, house building and the like still used there would be of great interest to me as a student of such things in the European past. Accompanied by Grith Lerche of the International Secretariat for Research on the History of Agricultural Implements in the Danish National Museum, I spent two weeks in New Guinea in September–October. Our guide was Ron Lampert, one of Golson's colleagues who, for a few days, conducted us round contemporary settlements and archaeological sites in the upper Wahgi valley, during which period Grith Lerche and I made the Kindeng, Mendim, Kum and Mount Hagen records described in the text. Our most important visit however was to Dr Nancy Bowers, who arranged for the reception of Grith Lerche, Ron Lampert and myself into the community of Alipe, where she was carrying out the ethnographic research which resulted in her doctoral dissertation *The Ascending Grasslands* (1968).

Alipe is a rather dispersed community on the homesteading pattern characteristic of the Western Highlands of Papua New Guinea. It is in the Kaugel valley at the eastern foot of Mount Giluwe, an impressive extinct volcano which is the second highest mountain in the country (4368 m). The Hagen Range (3800 m at its highest) is to the north-north-east, and the Mount Hagen township and the upper Wahgi valley are about 60 km away north-eastwards by road over the southern slopes of that range. At somewhat above 2000 m altitude, Alipe is beyond the limits at which most of the tropical crops on which New Guinea agriculture is based will produce. At this altitude yams cannot be grown, banana and sugar cane are rare, and the variety of green leafy or stem vegetables is reduced; only taro of the ancient starchy staples grows well and this is overshadowed by the more recent sweet potato of American ancestry (Bowers, 1971, p. 22). During our five-day stay at Alipe, we recorded houses and the operation of traditional gardening spades, and an old man demonstrated the use of a stone axe. Although two technical papers have already been published as a result of this 1968 visit (Lerche, 1969; Lerche and Steensberg, 1973), most of the material is gathered together in this volume.

In January 1971 I was again able to visit New Guinea. I had been asked to act as a consultant for UNESCO which was initiating an Oceanic Cultures Programme and I attended a meeting in Canberra at which this was discussed. The Australian National Commission for UNESCO made it possible for me to spend a few days in Papua New Guinea, then still under Australian administration, in connection with the consultancy. In the course of this I met Michael Somare, now Prime Minister of independent Papua New Guinea. Somare was at that time President of the Board of Trustees of the Public Museum and Art Gallery of Papua New Guinea, and he continues in the same capacity for what is today the National Museum, an important agency in the field of traditional culture, superbly housed in a striking new building. I spent five days visiting villages around Wewak and Maprik and along the Sepik river. The records from Aupik in the present volume derive from this visit.

Already I had conceived the idea of the present book: indeed I presented some of my ideas in a small but well-illustrated book in Danish (Steensberg, 1973), which received surprisingly wide and favourable notice and many requests to publish an edition in English. However, it seemed to me necessary to get back to New Guinea for a longer period to make further observations on matters that had interested me on my previous visits. This related particularly to the use of stone axes and adzes and the tools of tillage. When finally I was able to make this visit in 1975, the time was propitious for my purposes.

Peter White of the University of Sydney had just completed a detailed study of stone tool technology in all its aspects amongst the Duna of the far western reaches of the Southern Highlands, who only began receiving metal tools about 1960 (cf. White and Thomas, 1972; White *et al.*, 1977). In this work White co-operated with a social anthropologist, Nicholas Modjeska of Macquarie University, Sydney, who speaks the language and whose unpublished doctoral dissertation (for The Australian National University) is called *Production among the Duna* (1977). On White's advice I decided to make the Duna the focus of my projected study of tree-felling with stone tools. He gave me my introduction to his main informant, Irari Hipuya, who also became mine. All the records, not only of tree-felling but of housing, fencing and gardening in Puya-kira'go, Yuku and Yaku were made during the twelve days I spent in this area in September 1975.

My work on traditional gardening tools was concentrated in the upper Wahgi valley where Jack Golson was conducting his fourth season of archaeological investigations into the history of New Guinea Highlands agriculture, in the swamp deposits of the Kuk Tea Research Station, 13 km east of Mount Hagen township at 1550 m altitude. In every respect the situation in the upper Wahgi valley contrasted with that at Puya-kira'go. In the Wahgi valley, contact goes back to the first effective entry of Europeans into the Highlands in 1933 and has become more intensive over the past 25 years. In addition, in contrast to the forest edge situation at Puya-kira'go with its mosaic of secondary growth, scrub and grassland, the Wahgi valley is an area of extensive grassland where millennia of agricultural activities have almost totally removed any kind of forest and where the landscape is relieved by groves of planted casuarina and bamboo. Finally, in contrast to the situation at Alipe, we are here at an altitude where all the major crops of New Guinea agriculture can be grown, besides recently introduced cash crops like coffee. The observations on gardening tools and tree-felling from Kuk and Mbukl were made during my twenty days stay with Golson during September and October 1975.

So although my three visits to Papua New Guinea were short they provided the basic information for this book.

Copenhagen
April, 1980

Axel Steensberg

Acknowledgements

First and foremost I record my debt to the people of Papua New Guinea, whose traditional skills are recorded in this book. Specifically I would like to thank the Prime Minister of Papua New Guinea, Michael Somare, who took a personal interest in my work; to the other State authorities through which my research was approved; and to the people who helped me carry it out, particularly those associated with my work in 1975—Irari Hipuya, Dogore Dipaya and Dundaki of Puya-kira'go, with my interpreter Balana Pelake of Kopiago who accompanied me there, and, in the Hagen district, Ul, El and Ru.

Ulli Beier, Director of the Institute of Papua New Guinea Studies in Port Moresby, helped me with the formalities of my 1975 visit, and in other ways, while Andrew Strathern, then Professor of Anthropology and Sociology at the University of Papua New Guinea and now Professor of Anthropology at University College, London, gave my project the sponsorship of his Department. In the field during 1975 Peter Turner, Patrol Officer at Kopiago, provided invaluable assistance, as at Hagen did Douglas Grace, Manager of the Kuk Tea Research Station, Alan McGrigor, Principal of the Highlands Agricultural College, N. A. Robinson of the same institution, and Denis and Alison McEncroe of Mount Hagen township.

I owe a special debt to Peter White, Department of Anthropology, University of Sydney, who on the basis of his own fieldwork amongst the Duna, recommended Puya-kira'go as an appropriate place for my proposed studies of 1975 on the use of stone axes in tree-felling, and provided me with an introduction to his informant, Irari Hipuya, who became my own. Similarly at Mount Hagen, Jack Golson, Professor of Prehistory in the Research School of Pacific Studies at the Australian National University, and his colleagues Philip Hughes and Klim Gollan, made facilities available and provided assistance for my work on traditional gardening tools during their 1975 archaeological investigations into the history of Highlands agriculture at Kuk Tea Research Station.

I must not forget to thank Ron Lampert, then of Golson's Department and now at the Australian Museum, Sydney, who accompanied me into the field in 1968 and Nancy Bowers, at the time on the staff of the Department of Anthropology, Duke University and now at the University of Auckland, who made my work in the Kaugel Valley possible on that visit.

Many other people in Papua New Guinea gave me the benefit of their advice and discussions on one or more of my visits. At the University of Papua New Guinea these were Ralph Bulmer, Professor of Anthropology and Sociology during my first visit, and his wife Susan, Gerard Ward, Professor of Geography at the same time, William C. Clarke who was Professor of Geography during my 1975 visit, and three of the lecturing staff of the University, Mary-Jane Mountain (Prehistory, 1975), Edgar Waters (History) and Father Anthony Ruhan S. J. (History of Science and Technology). I am also indebted to Brian Egloff, who was Curator of Archaeology at the National Museum in 1975.

After my 1975 season I was able to pay a most useful visit to Australia where I had valuable discussions at The Australian National University with John Mulvaney and Anthony Forge, respectively Professor of Prehistory and of Anthropology in the Department of Prehistory and Anthropology, School of General Studies of that University, Ian Hughes of the Human Sciences Programme, and Jim Allen of the Department of Prehistory, Research School of Pacific Studies. In Sydney I talked with Jim Specht, Curator of Anthropology at The Australian Museum, and in Bandung, on my return trip home, with Ian Glover of the Institute of Archaeology, University of London. I also wish to express my warm thanks to Grith Lerche, my companion on the 1968 visit to New Guinea and Australia. In Copenhagen Svend Jørgensen and J. Troels-Smith of the Danish National Museum gave me the benefit of their experience of tree-felling with flint axes. I have also learned from H. C. Dosedla, Hohenheim University, Stuttgart, who has done fieldwork in the Hagen area.

As regards the manuscript which I prepared on the basis of my New Guinea researches, Jack Golson worked with me on its revision and re-arrangement, while Fiona Steele laboured painstakingly to render my Danish expression more understandable to the English-speaking reader. Nancy Bowers, Andrew Strathern, Peter White and Nicholas Modjesca were most helpful in answering queries, especially on the rendering of vernacular names and terms. I here formally thank all the persons and institutions mentioned individually in the text for permission to reproduce illustrations and quotations. All other illustrations are the author's.

Last but not least my sincere thanks go to the Danish Research Council for the Humanities which provided travel funds for my 1975 fieldwork and supported the processing of research data from it, and to the "Bequest to the Memory of Brewer J. C. Jacobsen" which contributed to my stay in Papua New Guinea in the same year.

In conclusion I should like to emphasize that I alone am responsible for any errors that are to be found in the text that follows.

Contents

List of Figures

Figure

1 | Clearing The Ground

Before the introduction of iron implements, tools for clearing land for cultivation were made of naturally occurring materials such as stone, wood and bamboo. It is these that we will be primarily concerned with here because their manufacture and use is of the utmost importance to an understanding of early cultivation.

Over the years since the early 1930s when Europeans first entered the New Guinea Highlands, some traditional forms have been replaced by their metal equivalents. In most places in Papua New Guinea, however, there are men still alive who grew up in the years before metal became common, and in some remote places where European tools began to displace traditional ones more recently, knowledge of the latter is vivid and widespread. The information that follows was obtained largely in two places that stand at the two extremes of this spectrum—at the community at Kuk near Mount Hagen, a region where people have been in continuous contact with Europeans since 1933, and at Puya-kira'go, out from Lake Kopiago in the Southern Highlands Province, where Europeans did not effectively penetrate until the late 1950s.

1. Tree-felling Tools: Axes and Adzes

Spiritual and material requirements were intimately interwoven in the evaluation of the stone adze and the axe blades produced in the New Guinea Highlands. Blades made of the right material were considered valuable not only because of their suitability as working tools, but also because they were taken out of rocks where they were believed to be born and to reproduce themselves.

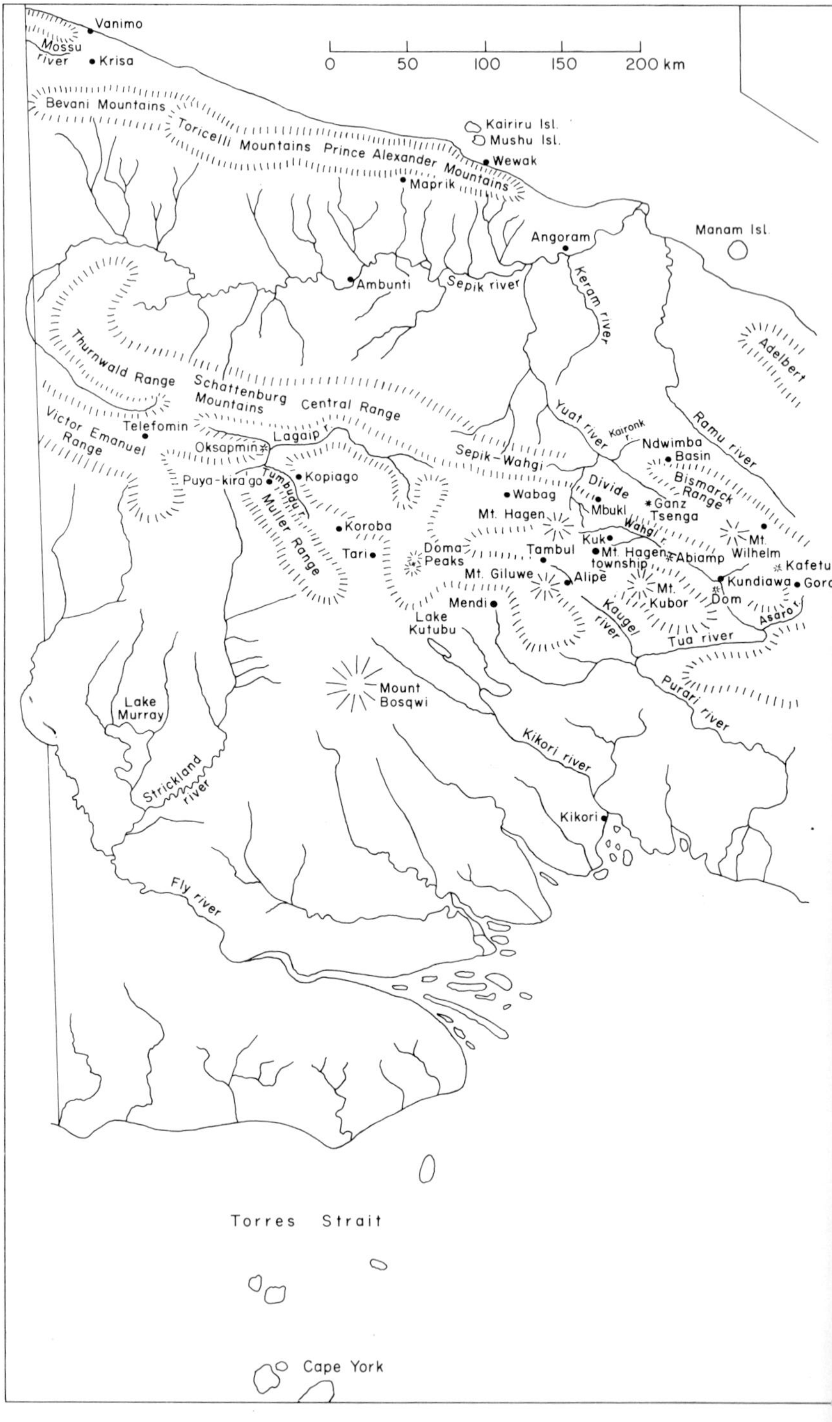

Fig. 1. Map of the localities of Papua New Guinea mentioned in the text.

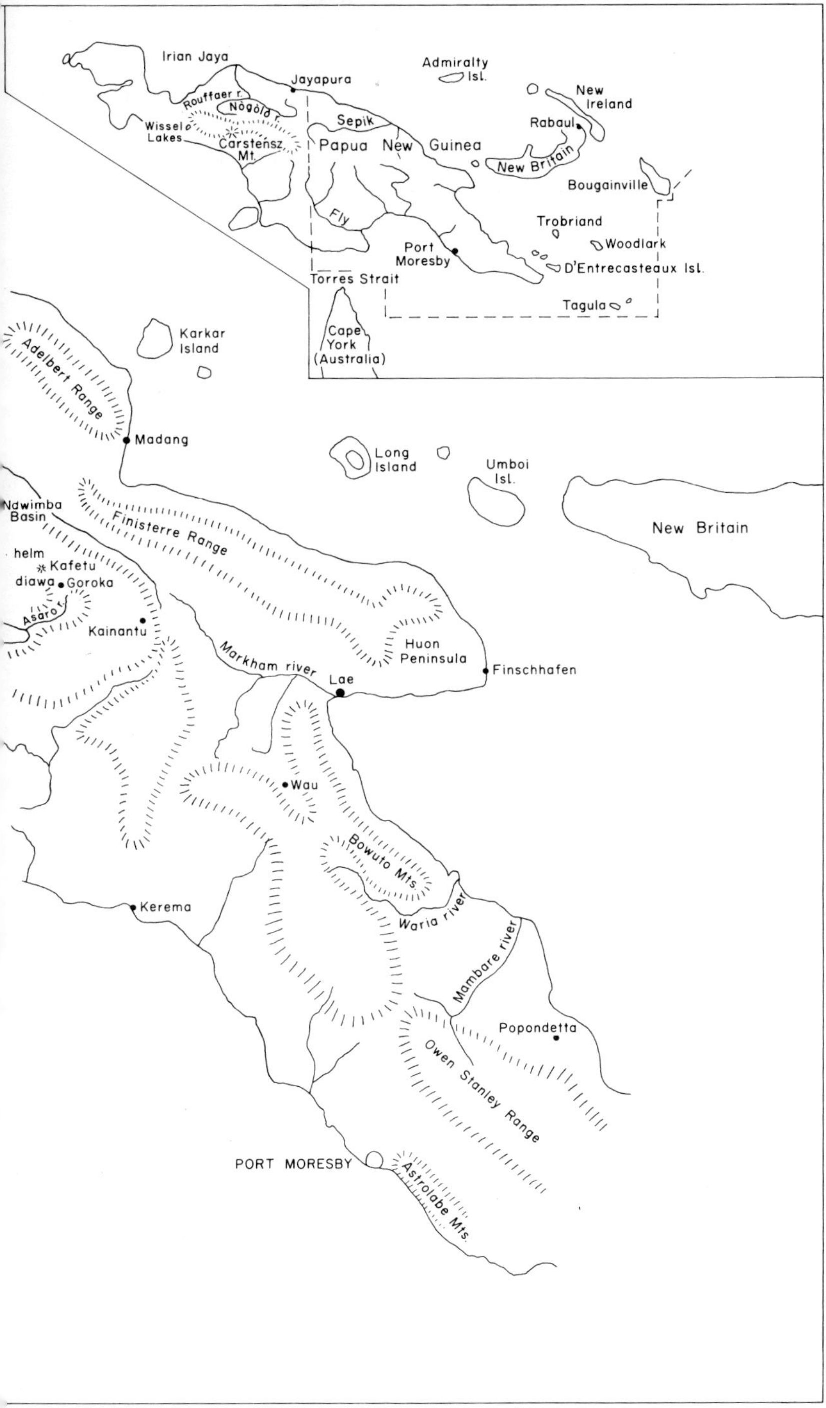

Irian Jaya
Jayapura
Admiralty Isl.
New Ireland
Rabaul
Rouffaer r.
Nogolo r.
Wissel Lakes
Carstensz Mt.
Sepik
Papua New Guinea
New Britain
Bougainville
Fly
Trobriand
Woodlark
Port Moresby
D'Entrecasteaux Isl.
Torres Strait
Tagula
Cape York (Australia)
Karkar Island
Adelbert Range
Madang
Long Island
Umboi Isl.
New Britain
Nawimba Basin
Finisterre Range
helm
Kafetu
diawa
Goroka
Asaro r.
Kainantu
Huon Peninsula
Markham river
Lae
Finschhafen
Wau
Bowuto Mts
Waria river
Kerema
Mambare river
Popondetta
Owen Stanley Range
PORT MORESBY
Astrolabe Mts.

In some areas the most valued blades were made of quarried materials thought to be endowed with a special strength or power. Some of them were used in cult performances and for ceremonial exchange, and on occasion they could be greased with fat from a pig offering in order to mobilize and strengthen their power. They might also be rubbed with sap from trees and plants or even with a man's semen in order to encourage fertility. And though axe blades made of stone are of less practical value than modern steel axes they are still thought to be superior in cult value (Dosedla, 1975). Such a treatment of holy stones can be compared with the common practice in northern Europe of smearing cup-stones with fat, for example the Swedish offerstones (Hammarstedt, 1914–1915).

Important Highlands quarries were situated between the Sepik-Wahgi Divide and the Kubor Range to the west, and Dom and Kafetu near Kundiawa and Goroka to the east (Hughes, 1977) (see Fig. 1). In former times blades from these quarries seem to have been traded right over to the border of what is now Irian Jaya. But there are quarries in the Central Range from which until recently the people of Puya-kira'go acquired their stone blades in exchange for bows. The local name around Lake Kopiago for an axe is *ayu*. The name used by the Hageners is *rui* and by the Kuman speakers of the Middle Wahgi is *di*, while in Neo-Melanesian (Pidgin) the axe is called *tamiok*, derived from "tomahawk".

In Irian Jaya there were important quarries in the upper Nògòlò river area, halfway between the Carstensz (now Jaya) and Wilhelmina (now Trikora) peaks, from where, according to Le Roux (1948, p. 323), axe blades were traded hundreds of kilometres westwards and eastwards. One of his informants demonstrated how to make an axe blade out of a water-rolled boulder on the banks of the Nògòlò river. He brought a boulder the size of a man's head and placed it on a pile of burning fuel. When the stone was heated sufficiently, cold water was poured on it through a bamboo pipe so that it cracked into large pieces. The most convenient piece was selected from the fire with a pair of bamboo tongs, returned to the fire for another more superficial heating, taken out again with the tongs and sprayed with water. Then, using a conical piece of an old axe blade, the man chipped it until he obtained the intended shape. Finally it was ground on a hard sandstone (Le Roux, 1948, p. 425).

In 1971 I saw a heavy sandstone block full of deep concavities in which men formerly sharpened their stone blades. This block (Fig. 2), which is mounted in a prominent position in the village of Aupik

Fig. 2. Heavy stone block in Aupik village, East Sepik Province, with about 30 grooves in which stone blades have been sharpened, 1971.

near Maprik in the Wewak Sub-Province, is not unlike whetstones found in northern Europe, used for the same purpose in the Neolithic (Steensberg, 1975).

2. Hafting Stone Blades at Puya-kira'go and Kuk

In 1975 Dogore Dipaia was between 40 and 50 years old. He must have been about 30 when the first steel tools appeared in his district, and so had had at least ten years experience in his youth of the manufacture and use of stone tools. Because of this, Irari Hipuya, my main contact, asked Dogore to demonstrate for me the hafting of a stone blade and subsequently its use in tree-felling.

One morning Dogore sent one of his friends into the rainforest to cut some rattan suitable for lashing an old stone blade to a haft he was going to make. Figure 3 shows Dogore's friend with the rattan, *kelewa* in the local language. He split one end of the rattan with a splinter of chert, then with one part of the split end gripped in his teeth, he pulled the other part with one hand, using the chert flake in his other hand to split the rattan along its length. Splinters were easily made from flaky boulders stored along the wall of the house. They were simply detached by using another stone as a hammer and were thrown away as soon as the work was finished. A professional archaeologist might not realize that such rough flakes had been used as tools; he would probably classify them as rubbish, waste from a Neolithic workshop.

The rattan strip was then wound round the maker's big toe and scraped with a stone splinter until it was the proper thickness (Fig. 4). As one part was finished, the strip was loosened and moved so that another stretch could be treated in the same way.

While this work was in progress, Dogore began to make the haft. This consisted of two parts, a separate *socket* of wood for the blade, and a *helve* made from a forked branch from the forest. The slender part of the fork was intended for the shaft or handle, while the part cut out of the stem was to be hollowed out on one side to make a bed for the socket and its blade. This was done in such a way that one side was higher than the other, producing a flange at an angle of approximately 90° to the base, against which the socket would back. The reason for this is that when the cutting edge of the blade cuts into a tree trunk with its left face against the stem, this upright flange will be on the right-hand side of the axeman away from the trunk, to take the shock and prevent the socket turning over to the right and breaking its lashing. When subsequently Dogore smoothed the handle with his bush knife, the groove for the socket could be photographed clearly (Fig. 5).

Subsequently Dogore carried the helve into the men's house and put its handle into the fire in order to make it flexible. Before it began to burn, he removed it from the fire and carried it to an old tree-stump outside the house, where he wedged it so that he could force the handle into the required shape (Fig. 6). After he had whetted the cutting edge of the blade on a hard sandstone, dipping it frequently in a small pool of water in front of him (Fig. 7), he

Fig. 3. Rattan collected from the forest to provide lashings for the hafting of a stone blade, Puya-kira'go, near Lake Kopiago, Southern Highlands, 1975.

Fig. 4. The rattan was split into strips which were then scraped with a splinter of stone until the correct thickness was obtained. The strip was wound round the big toe and held taut by the left hand.

Fig. 5. Dogore fashioning the helve from a forked branch with his steel bush knife. At the top he has prepared the seating for the socket in which the stone blade is to be mounted. The flange at the back will be to the axeman's right during use.

Fig. 6. After heating in a fire, the shaft of the helve was wedged in a tree-stump so that it could be forced into the required shape.

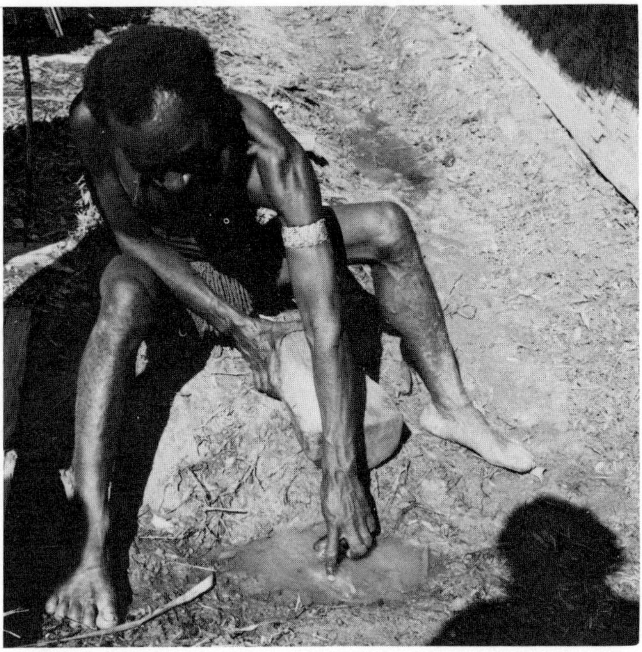

Fig. 7. Dogore dipping a stone blade in water during the process of sharpening it with a sandstone, Puya-kira'go, 1975.

Fig. 8. The socket for the stone blade was made of a shaped piece of wood that was split so that the seating for the blade could be hollowed out inside.

was ready to make the socket into which the blade was to be fitted.

He fashioned a piece of wood, tapering it at one end and making it elliptical in cross-section. This piece was not made of the heart of the wood but of one side of the branch so that when he split it from the broad end with his axe (Fig. 8) in preparation for the seating of the blade, the tree rings on the outside were not as abruptly curved as they would have been had he cut through the heartwood of the branch. Apparently this was done in order that the socket should not split so easily. The longitudinal crack was not made exactly in the middle of the elliptical billet but was displaced a little to the outside. Then Dogore placed the stone blade temporarily in the crack in order to calculate how to fit it in both planes. With a transverse cut he detached the narrow part near the pointed end so that the two pieces, one thicker than the other, were separated, and he hollowed them inside to make space for the stone blade.

When this had been completed, Dogore was ready to tie the blade firmly into its socket. He placed two short pieces of rattan in the cracks at the two narrow sides of the blade so that their ends protruded, and he then began to wrap the rattan crosswise around the socket, which required all his strength (Fig. 9). The protruding pieces of rattan were embedded below the last layer of lashing. Now he was able to fit the socket to the head of its helve, i.e. the part hollowed out for it as described above. First he put the two pieces together temporarily with their top ends level, then he cut about 3 cm off the top end of the helve-head and finished fashioning both the head and the handle. Only at that point could he begin the final stage of wrapping the blade, embedded in its socket, to the head of its helve. This was done by winding a lashing strip around the handle, securing it, and bending it over the socket crosswise on either side of the handle (Fig. 10).

It should be noted that the socket was turned so that the cutting edge of its blade was set at an angle of about 30° to the handle or shaft, the concavity of the hollow-ground blade, which was in fact fashioned as an adze, facing the helve. The angle of the longitudinal axis of the blade to the handle was about 70°.

Although Dogore used a sharp steel axe for the hafting of the stone blade, it took him more than half a day to finish the work. He spent from 9:00 to 10:30 manufacturing the socket. The cutting edge of the blade had not been damaged, so the whetting of it took only three quarters of an hour and he started wrapping the blade into its socket at 11:15. His friend was still sitting scraping

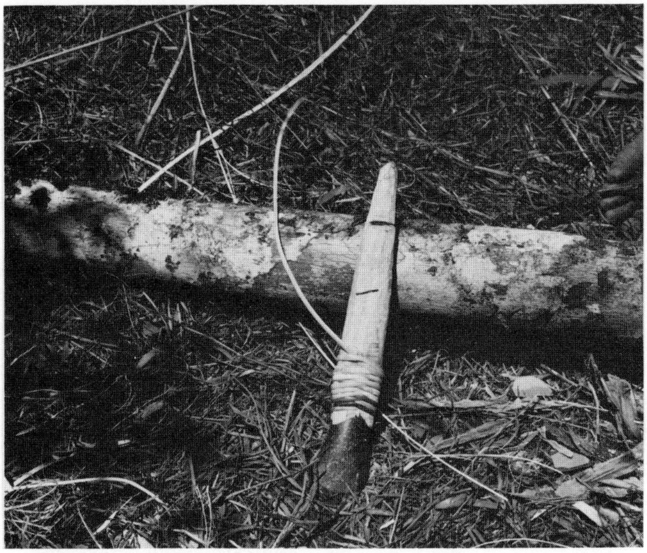

Fig. 9. The stone blade was bound into the socket with strips of rattan.

Fig. 10. The socket being attached to the helve with rattan lashings.

the rattan with a chert at 12:45 when Dogore gave the surface of the handle its last smoothing. When this was finished at 13:05, he put the handle into the fire and fashioned it again. The difficult job of tying the socket to the helve lasted from 13:25 to 13:55 and required considerable physical effort. Now and then Dogore added a new rattan strip either to the end of that which had been tied up, locking the two ends together by bending them over each other at a right angle, or he stabbed the new end into the wrapping texture and started afresh.

The hafting of this blade (Fig. 11) lasted in all about 5 hours, without any interruption. Had it been done using a stone tool and bamboo knife, the work would have lasted 2 or 3 days according to the makers' habitual working speed. The whetting of a damaged blade-edge could easily take most of a day.

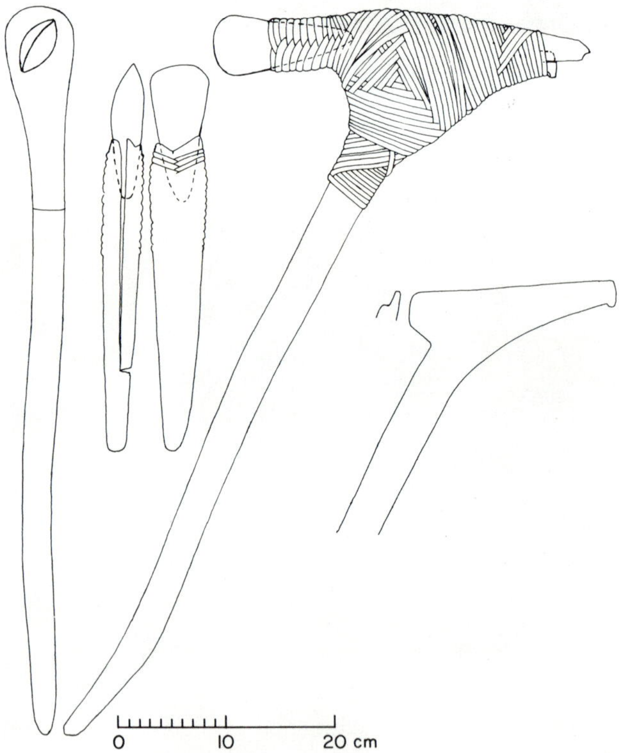

Fig. 11. The stone axe-adze from Puya-kira'go and its different elements. The view, far left, is end-on, showing the relationship between the line of the handle and the line of the cutting edge of the hafted blade. Similar views are also given in Figs 12–17. Drawn by the author.

Ul, a Hagener from the settlement of Mbukl on the Wahgi-Sepik Divide, was a man whose youth had been spent before European influence had made great changes in the traditional technology, and in 1975 at Kuk he was asked to haft an old stone blade in his traditional way. He started at 7:00 to prepare strips of bark for the lashing, having previously resharpened the blade. The strips were worked to and fro around a stick in order to loosen the fibres and separate out the less flexible parts allowing the removal of the bast from the bark elements with a flat wooden stick or knife. After this treatment the strips were each wound round a stick to prevent them getting tangled. The socket and helve had been made in advance and I was not told how long it took to make them. At 9:00 Ul began to lash the blade, which had been fitted into its socket, to the T-shaped helve. The ends of the lashing were joined by bending the broad end of the new piece round into a loop and twisting the narrow end of the piece to be lengthened around it making a kind of knot. In other cases the loose end was pushed in between the lashing with a wooden stick. Hafting was complete by 9:30.

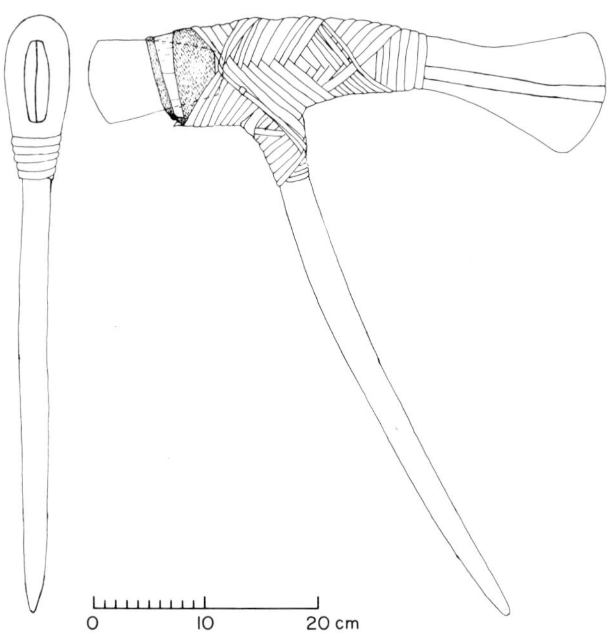

Fig. 12. Stone axe hafted in the Hagen fashion by Ul from Mbukl. The blade was an old one from Wapenamanda in Enga and Ul sharpened it with a facet along one side of the cutting edge. Drawn by the author.

This kind of lashing is not very durable in dry conditions. However, these implements, like other wooden tools, are usually kept in moist places, and eventually put into water in a ditch so that the hafting can absorb moisture and tighten up again. The top of the wooden socket of Ul's axe was shaped like an axe blade in the traditional fashion of the people of the Mount Hagen region; it was possibly made in this way in order to counterbalance the stone blade. One broad band was turned around the end of this wooden blade in order to fasten the whole lashing to the socket more securely; this is also a typical Hagen feature. The stone blade was hafted almost exactly in the same plane as the handle, but the handle was bent backwards at an angle of 110° to the longitudinal axis of the blade (Fig. 12), which is not a usual feature.

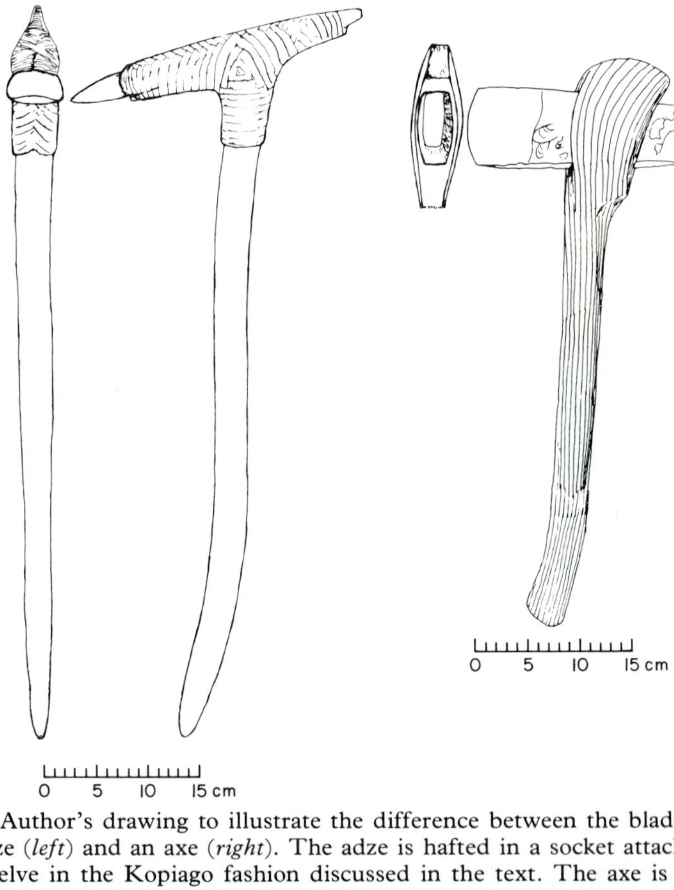

Fig. 13. Author's drawing to illustrate the difference between the blade settings of an adze (*left*) and an axe (*right*). The adze is hafted in a socket attached to an angled helve in the Kopiago fashion discussed in the text. The axe is hafted in the fashion of the Neolithic Danish Sigerslev axe in a perforated handle.

3. Styles of Hafting

These two tools are typical of what may be called a composite hafting style, widespread in the Highlands, using a socket and complex lashings. The socket gives an elastic connection between the blade and its haft that reduces the risk of breakage during heavy use. It also permits the rotation of the cutting edge of the blade anywhere between the true adze with the cutting edge at right angles to the shaft, and the true axe with the cutting edge parallel to the shaft (Fig. 13).

The Kuk implement hafted by Ul (Fig. 12) was to all intents and purposes a true axe. Another Hagener from Mbukl, El, had a tool of lighter construction mounted with a stone blade about 10 cm long, 4.5 cm wide and a little more than 1 cm thick (Fig. 14). It was

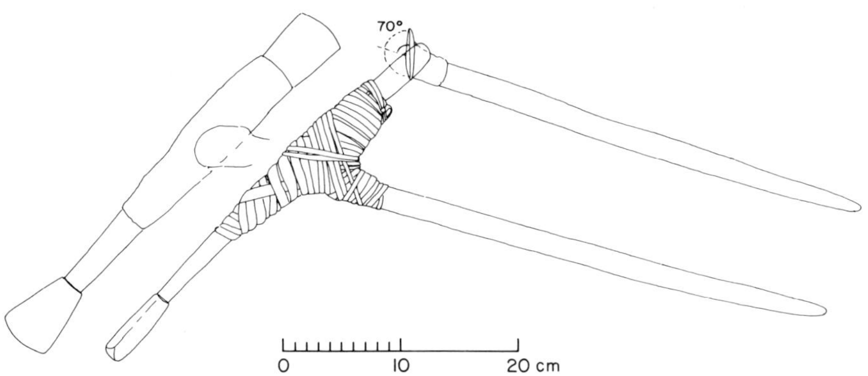

Fig. 14. Small axe-adze with a thin blade hafted with cutting edge at an angle of 70° to the handle. Used for finishing tools, hollowing out troughs etc. Drawn by the author.

furnished with a slender socket 32·5 cm long with an axe-like extension at the top like Ul's axe. Its handle was 43·5 cm long. The angle between the longitudinal axis of the blade and the handle was about 60° and the cutting edge was set at 70° to the median line of the handle; that is towards the adze end of the range between axe and adze. El had kept it for many years, because he found it useful, as a supplement to modern steel implements, axes or knives, in finishing light tools and hollowing out troughs.

Ul's axe corresponds with one seen in Alipe in the Kaugel valley south-west of Mount Hagen in 1968 (Steensberg 1973, Fig. 20 reproduced here as Fig. 15). The cutting edge was set at 7° to the

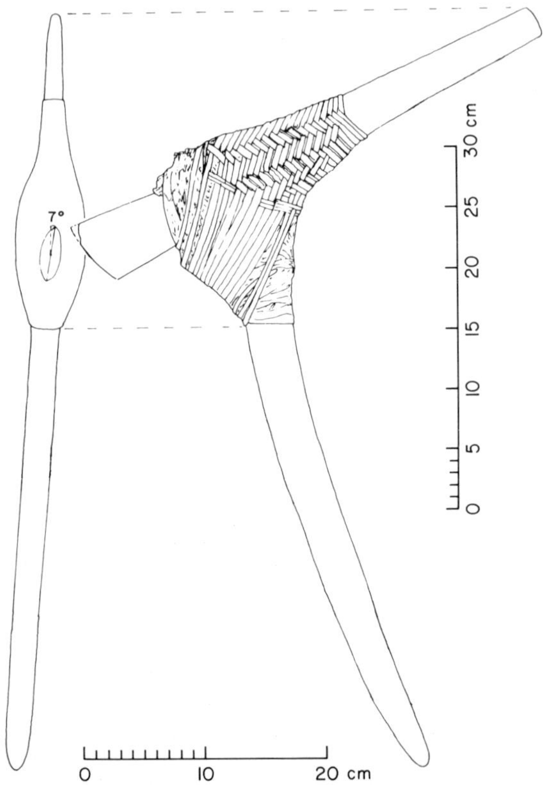

Fig. 15. Stone axe-adze from Alipe in the Kaugel valley, the use of which was demonstrated by Kupala in 1968. Drawn by the author.

median line of the handle, but the angle between the longitudinal axis of the blade and the shaft was 80°, compared with the 110° of Ul's axe. Note that the curve of the cutting edge of the blade is "skewed", i.e. not at 90° to the longitudinal axis of the blade. This phenomenon is also a common feature of stone blades from the Hagen area (Lampert, 1972) and its occurrence on blades in Irian Jaya was explained by Le Roux (1948, p. 414) as an attempt to avoid the blade striking the trunk with its inner corner rather than its full edge.

Another blade in its original haft came from Wapenamanda (Fig. 16). The shaft was almost at 90° to the longitudinal axis of the blade, the cutting edge of which was set at about 12° to the median line of the handle. The Puya-kira'go implement hafted by Dogore (Fig.

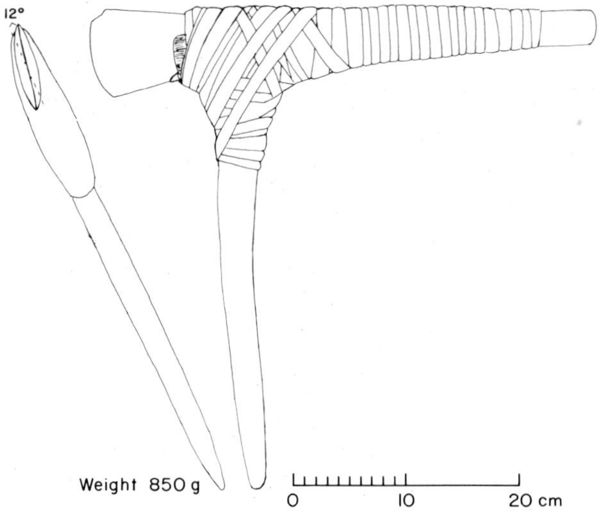

Fig. 16. Old axe-adze from Wapenamanda, Enga Sub-Province, belonging to N. A. Robinson of the Highlands Agricultural College, Mount Hagen (1975). The blade is set at an angle of 12° to the median line of the handle. Drawn by the author, and reproduced with Mr Robinson's permission.

11) had corresponding angles of 70° and 30°, and one acquired in the vicinity of the Lake Kopiago Patrol Post (Fig. 17) had angles of 77° and 42°. The latter two implements were hafted in the same manner, but the cutting edge of the Kopiago blade was not hollow-ground like the Puya-kira'go one.

The range of blade settings described, from about 0° for Ul's Hagen axe, through the 7° of the implement from Alipe, 12° from Wapenamanda, 30° from Puya-kira'go and 42° from Lake Kopiago, to El's example at 70°, suggests that the distinction between axe and adze may be rather an academic one. Because of this, archaeologists of New Guinea have adopted the neutral term axe-adze to describe stone blades, the setting of which may be variable.

There are, however, other hafting styles elsewhere in New Guinea which include true adzes and true axes. The axe blades of the southern areas of the island west from the Gulf of Papua were simply inserted in a perforation in the distal end of a rather straight helve. The perforation was made about a quarter of the way down the shaft in order to avoid cracking the wood, but in spite of this Le Roux (1950, p. 431) saw many shafts spoiled in this way. Moreover, the hafting end was often thickened to guard against this, although the thickening must have been very inconvenient when the axe was used

Fig. 17. A stone axe-adze from the neighbourhood of the Kopiago Patrol Post in the Southern Highlands. It is hafted in the same manner as Fig. 11, but the cutting edge is not hollow ground. Drawn by the author.

for splitting firewood, even when wedges were put into the crack to hold it open. The perforation for the blade in these tools was often made at an angle of 70° to the handle and Le Roux maintained that it should be conical in shape in order to fit the somewhat pointed butt of the blade.

There is a striking similarity between this method of hafting and that most common in specimens preserved from Neolithic Europe. The Danish specimens found in bogs (Blinkenberg, 1898; Becker, 1945, 1949) have a characteristic backward bend at the distal end (Fig. 13 right) and this is also typical of many of the specimens found in the Lake Dwellings of Switzerland (Müller-Beck, 1965), especially those from the Cortaillod culture. The Danish examples were the prototypes for axes used in tree-felling by Jørgensen and Troels-Smith in the early 1950s in Zealand and at Draved Wood, South Jutland, the latter the site of large-scale experiments over the years 1953–1955 into the full range of prehistoric techniques of swidden agriculture (Jørgensen, 1953, 1974; Steensberg, 1979).

One thing Jørgensen and Troels-Smith learnt in their experiments (Jørgensen, 1974), which Le Roux does not seem to have observed, was that while the narrow sides of the stone blade have to fit tightly in the perforation, the broader faces should not, if cracking is to be avoided. An obvious reason for this is that if the blade produces too much pressure against the two long sides of the perforation, which runs along the handle, it might open cracks along the grain. The narrow ends of the perforation on the other hand run across the grain and so are much less susceptible to cracking, especially if the blade fits tightly against them so that it cannot move up and down. Jørgensen and Troels-Smith also concluded that the angle of the longitudinal axis of the blade in relation to the shaft should not deviate more than about 10° from a right angle if the efficiency of the tool in hewing is not to be reduced. As we have seen, however, Le Roux commonly recorded an angle of 70° between the perforation for the blade and the handle in Irian Jaya.

Such perforated hafts could be provided with sockets or sleeves, as they were not only in New Guinea (Le Roux, 1948, pp. 441–442) but also in prehistoric Europe. Müller-Beck (1965, pp. 21, 35–38) describes a great variety of forms, some of them with wooden sleeves, others with antler sleeves. The majority seem to have been used for the trimming of wooden objects, and their stone blades are generally small. The advantage of fitting a perforated helve with a sleeve is that the cutting edge of the blade could be turned at different angles to the plane of its helve, as in the composite axe-adzes with angled hafts discussed for New Guinea above. This advantage was also characteristic of the celts of bronze and iron found in Schleswig and Denmark (Thrane, 1964, pp. 72–81).

Angled shafts made from forked branches have also been found in Europe similar to the ones described from the New Guinea Highlands. They date from the early Neolithic in Denmark, the so-called Ertebølle culture (Petersen, 1970, Fig. 18, and unpublished findings from Zealand and Jutland) and from the late Neolithic in Switzerland. In connection with the latter findings Müller-Beck (1965, p. 35) illustrates a similar style of adze from twelfth dynasty Egypt (*c.*1800 B.C.), though fitted with a blade of bronze and tied to the angled shaft with leather straps.

The distribution of these various forms of tools—axes, adzes and intermediate settings—is complex and interesting in New Guinea.

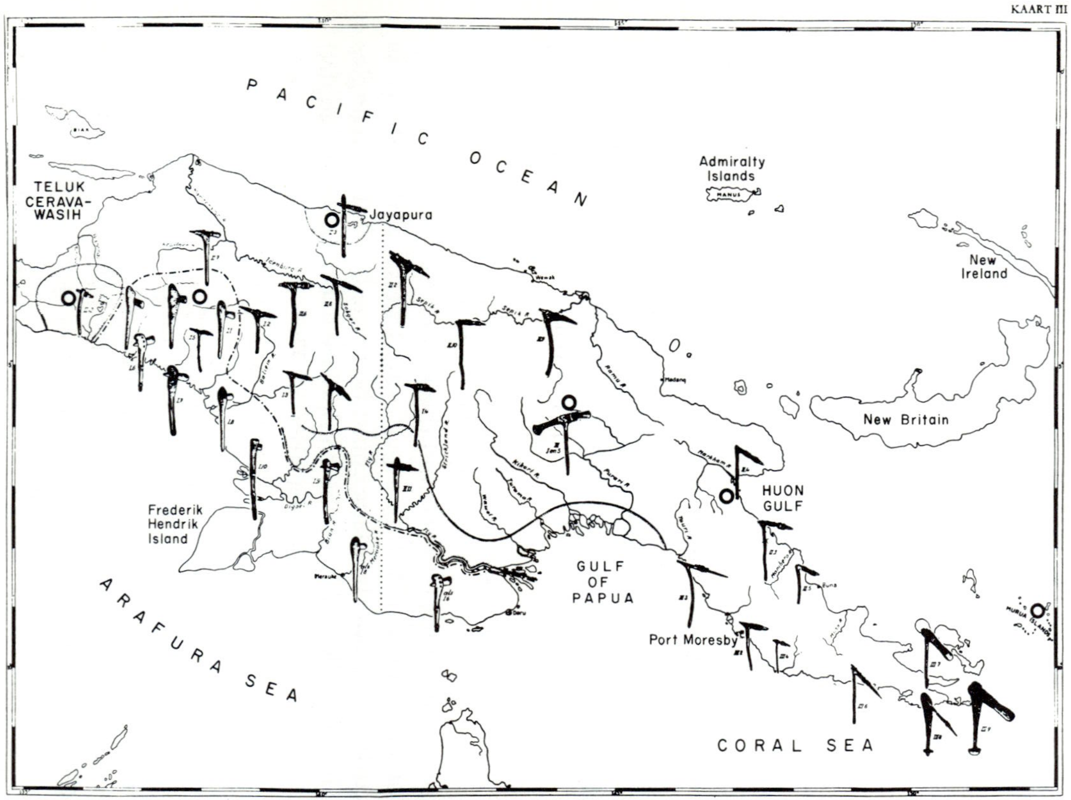

Fig. 18. Map of axe-adze types in New Guinea by Le Roux, mainly based on the collections and observations made during his stay on the island in the 1930s. The axes with straight handles are concentrated in the area facing the Arafura Sea from the Torres Strait westwards, including Frederik Hendrik Island. The circles indicate quarries for stone blades, and the numbers refer to places in a separate volume. After Le Roux (1948–50). With the permission of E. J. Brill, Publishers, Leiden.

A preliminary study was made by Le Roux (1950, Map III) whose map is reproduced here as Fig. 18. He only differentiated between the two extremes of axe and adze, not mentioning that sometimes the hafting was somewhere between the two.

A fuller study has recently been made by Crosby (1973, cf. Crosby, 1977). It is a meticulous and comprehensive analysis based on nearly a thousand complete axes and adzes of Melanesian origin in museums in the Australasian region and on information from a large number of ethnographic reports. The work only came to my attention at a late stage in the preparation of my manuscript, so that I could not use it to provide a general background for my own descriptions. These, however, are made from a functional point of view, where I try to put myself into the situation of the manufacturer and user, and learn their skills, while Crosby's is a stylistic approach, based on a study of the implements themselves supplemented by descriptions of their use from the ethnographic literature.

From her material she derives four traditions:

Tradition 1 (T1), direct hafting to a footed handle, is my angled haft without a socket.

Tradition 2 (T2), indirect hafting to a footed handle, is my angled haft with socket.

Tradition 3 (T3), direct hafting to a straight handle, is my straight haft with perforation for the blade and no socket.

Tradition 4 (T4), indirect hafting to a straight handle, is my straight haft with perforation for a blade inserted in a socket.

What is apparent is that the true adze form (Crosby's T1, my angled haft without a socket) is predominant in the Austronesian-speaking areas of south coastal Papua to the west and east of Port Moresby, in the inland mountains of the tail of Papua and in the highlands on both sides of the Papua New Guinea–Irian Jaya border (Crosby, 1973, Vol. I, pp. 267–274). The variant with socket (Crosby's T2) theoretically allowing rotation of the blade, dominates in the west of the island and along the northern coast, as well as in the large river valleys of the Sepik and Ramu and in the central highland zone of Papua New Guinea (Crosby, 1973, Vol. I, pp. 274–278). Around Lake Sentani near the north coast and the Paniai (Wissel) Lakes of the Irian Jaya highlands, as well as in the southern districts of the island from the Gulf of Papua westwards, axes with a straight handle (Crosby's T3 and T4) were used exclusively (Crosby, 1973, Vol. I, pp. 278–287). Her map of these distributions

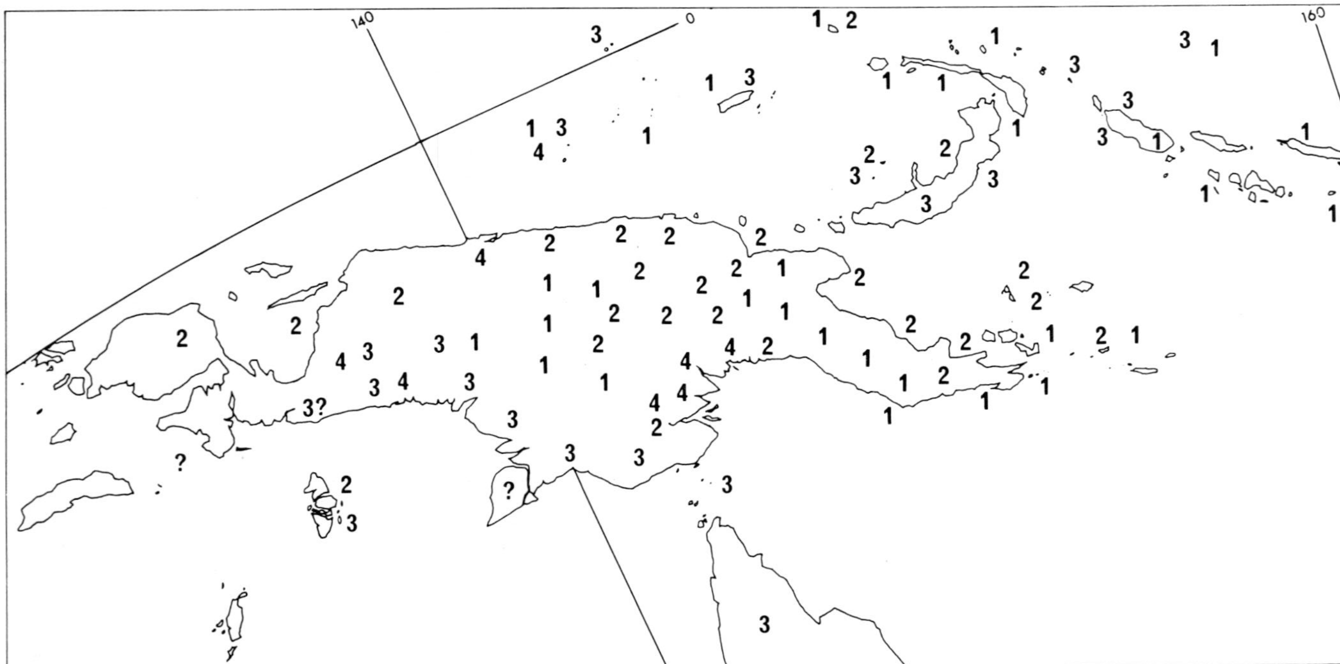

Fig. 19. Map of axe-adze types in Melanesia. 1 = Tradition T1, 2 = Tradition T2, 3 = Tradition T3, and 4 = Tradition T4. After Crosby (1973). Reproduced with permission.

(Crosby, 1973, Vol. I, p. 283, Map 27) is reproduced here as Fig. 19.

It should be appreciated that at times tools of different traditions were to be found side by side. Wirz (1924, p. 107) reports this for angle-hafted and straight-handled implements in the central parts of Irian Jaya, where axes with perforated handles were mainly used by women for the splitting of firewood.

No doubt the situation discussed above is a result of a complicated set of historical and ecological circumstances, and complex reasons must also underlie the situation in prehistoric Europe where different types of hafting have been found. I shall return to this question after I have reviewed the evidence of the use of stone tools in tree-felling. For the moment let us take the hafting of blades to angled shafts known both in New Guinea and prehistoric Europe. With hafting of this kind the recoil motion during use is caught by the lashing in a way comparable to the movement of an old naval gun in its gun-straps. One may ponder on the special challenges met by the lumbermen of highland New Guinea and temperate Europe when they independently invented an elastic method of hafting their stone blades.

There are of course other methods of achieving the same end. The use of a socket is one, known again both in New Guinea and prehistoric Europe. Also the actual technique of wielding the tool is very important (see later). Furthermore, the size and weight of an axe should be adapted to the stature of its owner, as Le Roux observed, though there are no simple rules as to the proportions (Le Roux, 1948, p. 432). In the Draved experiment two axes were particularly effective and remained completely undamaged. The first of these (No. 11–A1503 in Table I on p. 52) was furnished with a blade weighing 60 g and the total weight of the axe was 1122 g, which is less than the two implements from Puya-kira'go (Fig. 11) and Kuk (Fig. 12) before they dried out. The length of its helve was 58 cm. The second (M_1—b in Table I) had a total weight of 1178 g, and its helve was 62 cm long.

Figure 20 shows Jørgensen and Troels-Smith cutting a tree in Draved wood with these axes. I know of no corresponding data on perforated axes from New Guinea.

At Draved, Jørgensen and Troels-Smith found that the centre of gravity of an axe with a perforated helve was about 2–3 cm below the perforation, whereas the balance point of the angular hafted Puya-kira' go axe-adze lay in the angle between the shaft and the blade, and that of the Kuk axe lay in the angle between the shaft and

Fig. 20. Svend Jørgensen (*left*) and J. Troels-Smith (*right*) felling a tree with stone axes in Draved Wood, South Jutland, 1953. Photograph by Jens Steensberg.

expanded socket. Comparisons in this respect between newly made tools and specimens in museums are likely to be misleading, since the wooden parts of the latter will have lost weight while the stone blades will not, so that the calculations on this point by Crosby (1973, Vol. 1, e.g. p. 279) are probably of restricted value.

4. Felling Trees with Stone Implements

Le Roux (1978, p. 447) complained that very few references were available on the use of stone axes (and adzes), and the situation has changed very little since, probably because anthropologists themselves lack the necessary training and experience. Here it is interesting to note that although in the Draved experiment Troels-Smith and Jørgensen trained hard before they started tree-felling in earnest in the experimental area, it turned out to be something of an advantage to them that they had had relatively little experience previously of conventional felling with steel axes. They quickly learnt to avoid using the full swing of their arms and body whereas a professional lumberman who worked with them was inclined to lapse into his normal habit of using the whole of his arm and body, and consequently he broke more stone axes than expected. Although I was only involved directly in the last phase of the Draved experiment, namely, tilling, sowing and harvesting, I was experienced in the use of hand tools, and my observations of the tree-felling aspects provided a useful background for the New Guinea experiments to be described.

Shortly after the work at Draved was finished, I had a letter from Professor R. J. Forbes of Amsterdam relating to personal observations when he lived in Borneo in his youth. In his letter he wrote:

> In 1926 when I was working in a petroleum refinery on the east coast of Borneo, the Governor-General of the Dutch East Indies visited the place called Balik Papan and the delegations of the native Dajak tribes from the interior came to pay him homage. Some of these Dajaks still used stone axes, and so we arranged a tree-cutting contest between them and the Chinese wood-cutters who worked for us with modern tools to fell trees for the box factory of our refinery. The Dajaks had attached their stone axe-blades to a piece of bamboo which gave the axe a good flexibility and they chopped down a tree by forming a V-shaped incision. At each blow (from the elbow) they slightly turned the axe so as to cut along one of the two sides of the V-shaped incision. The blows were fairly rapid. The objects selected were two 15″ [*c.* 38 cm] trees of a kind called "tjemara" in the Malay language, a fairly tough pine-like timber. To our great astonishment the two Dajaks took only ten minutes to cut it down against 15 minutes for the Chinese with their modern American steel axes. One of the hold-ups of these modern wood-cutters was the extreme sharpness of their tool, which with the blow given from the shoulder drove the axe too far into the wood and caused some effort to extract the axe for the next blow. The Dajaks hammering the trunk with a series of sharp blows from the elbow and at much shorter intervals managed to drive their V-shaped incisions through the trunk much quicker.

From the experience of the Draved experiment, Jørgensen (1974) notes the same distinction between the steel axe with its sharp cutting edge and the stone blade with its wedge-like edge, which must operate by splitting the wood. Troels-Smith and he made a V-shaped incision to define the bottom of the cut they intended to make in the trunk and then chopped into the trunk above, splitting the wood down to the undercut incision. They then repeated the process, undercutting the incision and splitting from above, removing pieces *c.* 20 cm long at the beginning, which became progressively shorter as the incision went deeper into the trunk. Figure 21 shows the process almost complete. The undercutting did not produce marks as fine as the chopping from the top because of the awkward way of using the axe obliquely from below.

In *Seeberg Burgäschisee-Süd* (Müller-Beck, 1965) Schweingruber hypothesized about the method and effect of cutting a tree with a stone axe. However, he allows an angle of only 35–45° between the trunk and the median line of the axe blade when the axe is chopping from above (p. 69, Fig. 307). This is the angle seen on the right-hand side of the trunk in Fig. 21, but we should remember that this is the end-product. It seems likely that Schweingruber was misled by such prehistoric end-products to conclude that the angle represented by them was the angle of working. Jørgensen's diagram (1974, pl. XIII, 2) makes clear that the angle of working was about 55°, which corresponds better with the technique used in Puya-kira'go. The initial angle of the implement in the foreground of Fig. 20 seems to be even less acute than 55°.

The angle between the axe blade and the face of the trunk to be felled cannot be as acute as the incision visible after the tree has fallen, for the cutting edge must bite into the wood when it hits it and then gradually alter its direction, ultimately following the grain as it loses momentum. The initial angle of about 55° would also serve to prevent the massive upper part of a perforated haft, or the lashings of a composite one, being damaged by hitting the wood of the tree. But because of the thick lashing and the setting of the Puya-kira'go blade, the technique used here was slightly different from that practised at Kuk.

The tree felled by Dogore was a *dole*-tree 17 cm in diameter standing on a gentle slope in the rainforest. After clearing some scrub away, Dogore placed himself at an angle of about 45° to the face of the tree and started cutting it at chest height. This position was apparently the most convenient considering that the cutting edge of the blade was turned 30° from the plane of its helve. This meant that as the tool was swung downwards to strike the trunk, its cutting

Fig. 21. V-shaped incisions made with stone axes in a tree (*Alnus* No. 828) at Draved. Photograph by Svend Jørgensen.

Fig. 22. Dogore felling a tree in the bush at Puya-kira'go with an axe-adze (Fig. 11), 11 September 1975.

Fig. 23. Dogore lengthened his swing when the incision was nearing the centre of the tree.

edge was level. With the Wapenamanda implement in Fig. 16 the cutting edge of which is only at 12° to the median line of the handle, the lumberman would have stood in a position much closer to that when using a true axe.

The implement was swung in two slightly different ways, as can be seen in Figs 22 and 23. Initially Dogore swung it backwards and forwards over his right shoulder in short rapid strokes (Fig. 22), but when the incision was nearing the middle of the tree he used all his strength, swinging the implement further back and over the top of his head (Fig. 23). A flint axe would probably have been broken by such heavy strokes, and presumably also the thinner-bladed axe from Kuk would have been damaged. But Dogore had the benefit of his compact and less fragile blade with a curved edge lacking corners, as well as of a much stronger rattan lashing. He took full advantage of this, using the utmost force of his muscles, although his body moved only a little. His hands grasped the helve, his right hand between 30 and 21 cm from the tip of the handle, his left between 21 and 12 cm from the tip.

Afterwards Dogore made another incision on the opposite side of the trunk, his position being slightly higher because the tree stood on a slope. He arranged it so that the tree fell uphill in order that the camera could operate in a convenient position, but normally a lumberman would fell downhill. The tree was felled in 10 minutes, including a break of 3 minutes' duration when the interpreter and I tried our hand. The effect of our attempts was minimal, partly because I, for my part, placed my hands too far from each other.

Figure 24 shows the nature of the cut made by Dogore. The incision was made exclusively from above, in contrast to the undercut V-shaped incision used at Draved, and produced a very jagged and irregular lower part. Figure 25 illustrates the different end-product on a beech tree from Draved (Jørgensen, 1974, pl. XII). Figure 25a, the upper part of the Draved incision, shows how different the marks made by the flint axe are from those of the Puya-kira'go blade in Fig. 24. Figure 25b shows the lower face of the Draved incision, with a mixture of cuts and splinters.

Some of the characteristic wood chips from Puya-kira'go are drawn in Fig. 26. Each of them is shown from the front (left-hand drawing in each group), from the left side (middle drawing) and from the back (right-hand drawing). At the bottom of each front view are the clean marks made by the front of the blade on the timber during the cutting of a previous chip. The frayed border at the top

Fig. 24. The marks made by Dogore's axe-adze resemble the bites of a rodent. Note how the incision was made exclusively from above.

was made while the chip shown was being detached from the trunk. The cutting edge of the blade entered the trunk at an angle of about 55° and, being wedge-shaped, its back tore the top of the chip away from the trunk. Figure 26 (top middle, I-III) illustrates how wood chips with these characteristics were produced. The arrows indicate how, after entering the trunk at an angle, the blade turned almost parallel with it. Unfortunately, none of the bottom chips of the Puya-kira'go felling were collected. But most of them must have been splintered at the bottom as well as at the top because Dogore often broke them off by hand. In other cases he would loosen them with a cut of his blade.

In his study of the material from Burgäschisee-Süd, Müller-Beck (1965, Figs 6 and 275–276, p. 131) suggested how the Lake Dwellers felled trees. His reconstruction was based on fossil wood chips, one of which he drew full scale, 5 cm long, 1·7 cm wide and 0·6 cm thick. This is reproduced here as Fig. 27, left. He supposed that the cutting started from below, short chips being successively

(a)

(b)

Fig. 25. An oak felled during the experiments at Draved Wood. (a) shows the upper part of the cut (b) the lower part. Some of the chips in (b) had been broken (cf. Fig. 24) though most of them were undercut. After Jørgensen (1974).

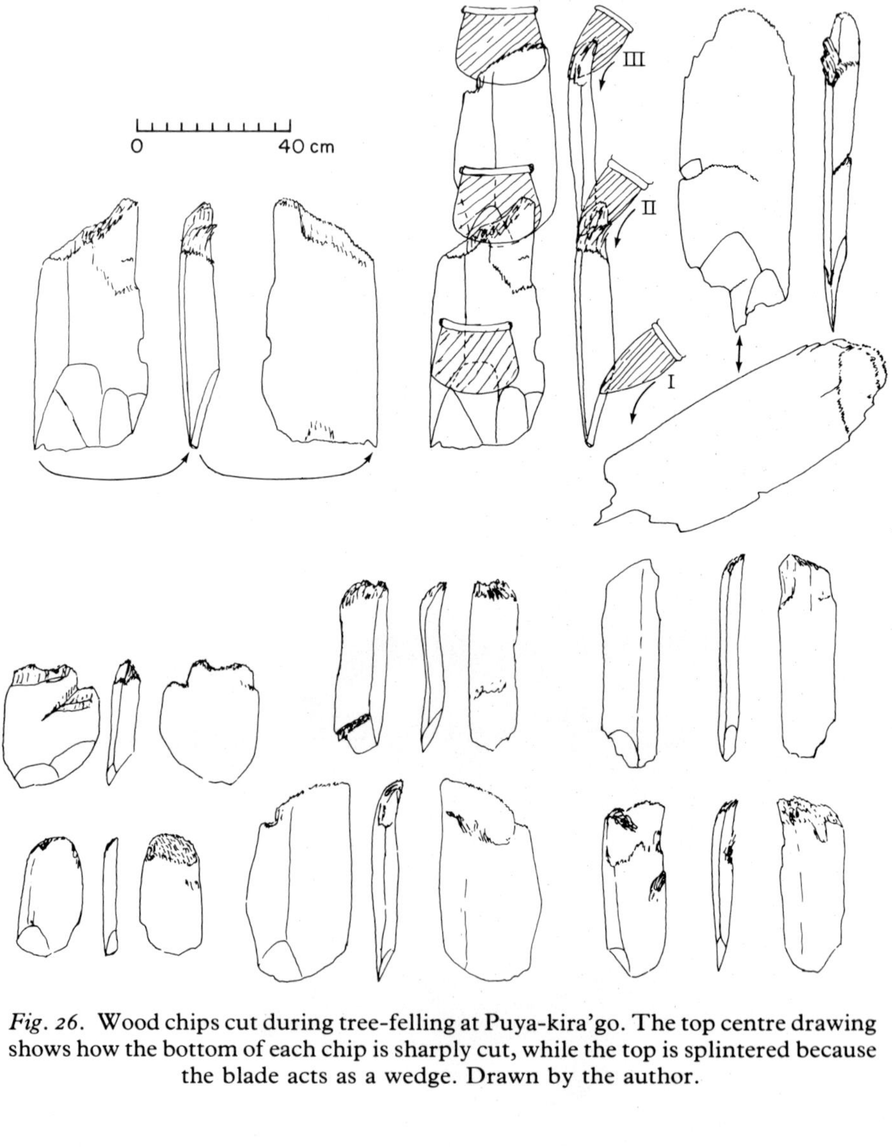

Fig. 26. Wood chips cut during tree-felling at Puya-kira'go. The top centre drawing shows how the bottom of each chip is sharply cut, while the top is splintered because the blade acts as a wedge. Drawn by the author.

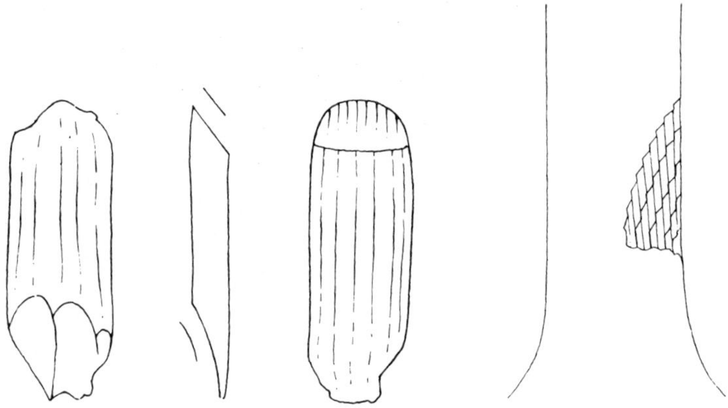

Fig. 27. Wood chip from Burgäschisee-Süd, Switzerland (Cortaillod culture) and (*right*) the way Müller-Beck suggested they were produced. After Müller-Beck (1965). With permission.

detached row after row upwards, the rows becoming shorter and shorter until the incision was deep enough (Fig. 27, right). He may well be correct in suggesting that the chipping started below and moved upwards, as it did at Puya-kira'go where the chips produced were also rather small. However, a lumberman is not likely to have worked in such a schematic way as Müller-Beck supposes. In the Kuk demonstration described below, though Ul tended to chop from the bottom upwards, he produced long chips. In the Draved experiments, Troels-Smith and Jørgensen cut full-length chips *downwards* from the top of the cut to the undercut incision.

The chip reproduced after Müller-Beck in Fig. 27, left, has the top looking as regular as the bottom. In his Draved report, Jørgensen (1974) described another fossil chip from the Swiss Lake Dwellings very similar in shape, 7 cm long, 5 cm wide and 1·2 cm thick. However, when he saw the recent Puya-kira'go wood chips with their frayed tops, he admitted that fossil wood chips deposited in a swamp for thousands of years could easily have lost significant details like this, frayed edges having decomposed and only more substantial parts being preserved.

Another man from Puya-kira'go, Yopa, who was about 40 years old and thus likely to have had 10–15 years experience with stone tools in his youth, felled a 9 m high tree in less than 5 minutes, using Dogore's axe and the same technique. The trunk was 13 cm thick and the stump he left was 97 cm high. This demonstration was filmed with an 8 mm Super camera: the original is with the

International Secretariat for Research on the History of Agricultural Implements, Copenhagen* and there is a copy in the Institute of Papua New Guinea Studies in Port Moresby. This tree was felled downslope.

The Hagener Ul, who assembled the axe shown in Fig. 12 and described above, demonstrated its use in felling a gum tree at Kuk Tea Research Station near Mount Hagen. Ul started by making two incisions through the bark around the circumference of the tree about 40 cm apart. He then made a vertical cut between the two horizontal incisions and prised the bark from the stem with the pointed end of his axe handle. This would seem to be the reason why Hagen axe handles are normally pointed. We may note that the shaft of the Wapenamanda implement (Fig. 16) from the Enga Province west of Mount Hagen was pointed in the same way.

Then he started cutting rather long chips, and in contrast to the practice used by Dogore, Ul always undercut the chips as seen in Fig. 28. This was probably due to the fact that his axe blade was like the Danish Neolithic flint axes used in the Draved experiment, while the Puya-kira'go axe was thicker so it could break the chips off without the necessity of undercutting them. But neither of the two men in New Guinea produced such a V-shaped lower incision as practised in Draved.

The tree felled by Ul was nearly 13 m high and the circumference where it was cut at a height of 65 cm was 67·5 cm. The felling took 13 minutes and 20 seconds, and Ul commented that the blade of his axe, which as mentioned previously weighed in a moist condition 1250 g, including the haft, was not quite heavy enough. Another day Ul felled a 14 m high gum tree with a circumference of 58 cm in 9 minutes of which about 1 minute was spent in drinking water. The stump left was about 65–70 cm high. These demonstrations were also filmed.

It is interesting to compare these results with the felling of a similar tree with a steel axe. El cut down a gum tree about 12 m high with a circumference of 59 cm in 2 minutes and 12 seconds with his steel axe, the stump left being only 57 cm high, i.e. about four times faster than with the stone axe. As expected his technique was different from Ul's because he used the full length of his arms in the sweep. But the angle of cutting was very much like that with the stone axe.

Some days before, both men had been in the mountain forest near

* c/o Danish National Museum, Brede, 2800 Copenhagen-Lyngby.

† P.O Box 1432, Boroko PNG.

Fig. 28. Ul from Mbukl ringbarked this gum tree at Kuk Tea Research Station near Mount Hagen before he began felling it with the stone axe (Fig. 12).

their home at Mbukl in order to cut a *yakla* palm with a steel axe. This is a tough timber which was to be used in the manufacture of some wooden knives and paddle-shaped spades (see later). The palm was on a slippery slope, awkwardly surrounded by luxuriant vegetation which made it difficult to film. However, an 8 mm Super movie film was produced of this felling, and copies are kept in the same two institutions as mentioned before. The palm was about 8·75 m high, with a circumference of 51 cm. El and Ul each took his turn with the axe, and the operation lasted only 2 minutes and 20 seconds, leaving a stump of 65 cm height. During the cutting I observed that the steel axe was sometimes inclined to stick in the wood because its blade was thin and sharp, just as Forbes noted in his Borneo letter quoted above. After they had cut a 3 m long piece of the stem, a younger man, Ru, carried it on his shoulder through the dense rain-forest, forcing his way up the steep and slippery slope to the road. Such men are remarkably strong when they are in their prime.

In August 1976 when my equipment, including the two stone implements from Puya-kira'go and Kuk, had been returned to Copenhagen, I conducted a small experiment cutting a stump of an almost dead elder tree (*Sambucus nigra*) at a height of 69–83 cm. The tree had been killed by hydrochloric acid dropped around its base, but a few green shoots could still be seen. Apart from this the wood seemed to be completely dried out, which meant that it was less easy to cut than if it had been fresh. It would also have been more convenient if the incision could have been made 90–100 cm above the ground or slightly higher, because I am nearly 10 cm taller (174 cm) than the men of Puya-kira'go and Kuk. This tree measured 11 cm in diameter and was 35 cm in circumference.

The incision was made alternately with the two New Guinea stone implements in order to reconsider the difference in their characteristics as cutting tools. The cutting time was not measured because part of it was taken up by observations and photography. Probably because of my low inconvenient position the lashing of the Kuk axe broke, and the blade turned outwards with the narrow side on my right forcing its way through the slot of its socket. The cutting edge must have hit the trunk at the wrong angle with the inner corner first. This means that a broad blade is weaker in its hafting than a narrow one if its broad cutting edge does not meet the wood exactly at the proper angle. In this case the blade was not itself damaged because its hafting was the weaker part of the tool. The lashing may have become fragile during the ten month journey by ship from Port Moresby to Copenhagen. Although it had been put in water in order to absorb moisture before it was used, apparently this was not enough to rejuvenate the lashing material.

The implement from Puya-kira'go was more solidly hafted and its rattan lashing was stronger than the bast of the Kuk axe. The lashing would have been broken of course if the left side of the tool hit the trunk of the tree, so the cutting angle had to be about 55° to the vertical. But the bulky blade would not have been easily torn out of its socket by the cut itself because its cutting edge was narrow and curved with no prominent corners to get in the way. The long handle of the Puya-kira'go axe allowed heavier strokes, which counterbalanced the Kuk axe's advantage of having a wider cutting edge. The axe from Kuk had to be used with shorter, more rapid strokes.

Townsend (1969) studied the Sanio-Hiowe (initially spelt Heve) people living in the foothills adjacent to the Wogamus river, a southern tributary of the Sepik river about 100 km north of Puya-kira'go. Steel tools had began to trickle into the area by about 1960 but had not fully replaced any traditional tools until 1966, and in that year a large house was still built using stone adzes. In 1967 Townsend asked six men, aged between 18 and 30 and equally skilled in the use of stone and steel implements, each to cut ten trees selected to cover a representative range of varieties, sizes and hardness of wood. They used a stone adze for the first five and a steel axe for the remainder. The felling of the ten trees required on average about an hour, including coming and going, approximately the length of time a Hiowe man would normally work at such a task. The experiment was then extended to comprise in all 91 trees of 14 named varieties, 31 of them being *Burseraceae* (the balsam family) which is the most common type in this area. Forty-five of the trees felled, ranging in circumference from 14·3 to 111·8 cm, were cut with a steel axe, and the remaining 46, varying in circumference from 16·5 to 124·5 cm, with a stone adze. The adze became dull after cutting 34 trees, which required 3 hours and 49 minutes, and had to be resharpened, which took about an hour.

The Hiowe notched the tree on one side, and then they notched it on the opposite side until it fell. With the adze they first cut at the bottom of the notch and then at the top splitting off a chip of wood. Because of the overhead stroke required by the adze, the tree was cut higher above the ground than when using a steel axe. Townsend added that large trees were cut using an adze at heights between 1·83 m to 2·44 m above the ground because of the buttressed roots; and clearing around the tree, rebinding and adjusting the adze head as well as making a simple scaffold as a platform for the cutter to stand on were all included in the time required to fell the tree.

Townsend found that it took, on average, about 4·4 times longer to fell trees with stone adzes than with steel axes. This is not inconsistent with the results obtained in the Kuk experiment in which the time spent in cutting trees of 53–75 cm circumference with the stone axe compared with that of the steel axe was about three or four times as much.

Felling with stone adzes has also been described by Godelier (Godelier with Garanger, 1973, pp. 187–220) based on observations during 1968–1969 among the Baruya people inhabiting about a dozen villages and hamlets in the Sub-Province of Marawaka, between Wonenara and Menyamya, in the Eastern Highlands Province of Papua New Guinea. The experiment comprised 91 trees in all, and the techniques of felling were filmed by Jablonko, while Garanger subsequently described the tools and performed a technological analysis of the adzes (Godelier with Garanger, 1973, pp. 190–197). However, in his Fig. 1 the scale must be 10 cm not 20 cm and whether the cutting edge is hollow ground or not is not indicated. Also there must be an error in his Fig. 2 describing the relationship between the length of the handle (abscissa) and the total weight of the tool (ordinate). Apparently the total weight of Nos 2 and 4 is the same as the weight of the adze-heads alone, indicated in his Fig. 3, which causes the anomalies of those two in Fig. 4, as well as in Fig. 5.

In this area large trees with huge buttress roots were attacked from scaffolds similar to those constructed by the Hiowe, and in both cases adze-hafted tools seemed to have been preferred to tools hafted as axes. The reason for this may have lain in the overhead stroke of the adze in that it cut the tree higher off the ground, and also this predominantly vertical movement would cause less vibration of the scaffold. Adzes were certainly the most common tools for felling in most parts of the New Guinea rainforest, and it would be interesting to investigate this further because the composition of primary and secondary woods in the Neolithic of Europe suggests that there was experience here of a similar kind. In the Marawaka area, Godelier noted that the deepest incision was made from the side opposite to that in which the tree was intended to fall. For example, Godelier mentions that a tree of 43 cm diameter was cut from one side to a depth of 25 cm, and from the other side, to which the tree fell, to only 15 cm. The adze tore and splintered the wood more than it cut it. The cutting went on in a series of alternately short and long rhythmic strokes. The adze started its sweep at a point above the cutter's head and often he pressed his thumb on the shaft in order to produce extra pressure.

Godelier (Godelier with Garanger, 1973, pp. 200–212) attempted over a period of four days to measure the efficiency of tree felling with stone adzes as against steel axes. He ran into the same sort of difficulties in reviving the use of tools and techniques now no longer employed as I myself did in my effort to study the use of traditional spades in the Mount Hagen district, as described in Chapter 2. However, his general conclusion is that it takes $1\frac{1}{2}$–2 times as long to fell small trees with stone as compared with steel tools, and that when the tree is a metre or more upwards in circumference it begins to take 3–$4\frac{1}{2}$ times as long. The latter is inconsistent with Townsend's observation for the Hiowe, because his graphs (Townsend 1969, Fig. 4, p. 203) suggest that the superiority of steel over stone in terms of the time taken to fell a tree of given size decreased above a circumference of 71 cm.

Heider does not publish figures on the effectiveness of stone adzes among the Dugum Dani in Irian Jaya, but according to his illustrations the technique practised is similar to that of the Baruya (Heider, 1970, p. 276, pl. 50–56; Gardner and Heider, 1968, p. 28, Figs 45–46 and p. 46, Fig. 133).

When Le Roux visited the Rouffaer river area before the Second World War, he noted that if people had to fell trees of more than 75 cm in diameter they built 1–2 m high scaffolds in order to save labour because of the buttressed roots, though the largest trees were left standing, being simply ring-barked so that they would lose most of their leaves and thus not shade the gardens (Le Roux, 1948, p. 449). Often young trees were cut halfway through on one side only so that the wind would blow them down, while larger trees were cut all the way round and brought down by pulling at their branches. The tools were sometimes soaked in water before use in order to make the hafting stronger. Le Roux pointed to two features that have figured in our own observations. He noted that the cutting edge of the blade was not always parallel to the handle, what we have called "skewed". In his view this was to avoid the blade striking the trunk with its inner corner rather than its full edge. He also noted that one face of the stone blade was often more curved longitudinally than the other, so that they met the cutting edge at different angles. This meant that the stone blade did not stick in the cut as sometimes happens with a steel axe (Le Roux, 1948, p. 414).

According to Townsend's experiments among the Hiowe, the advantage of using steel axes instead of stone axes decreased slightly for trees of 71 cm circumference and upwards, whilst according to Godelier it was the opposite. However, the time required for cutting the timber for house building was not the principal factor involved in that operation (Townsend, 1969, p. 204); the reasons for the frequent choice of secondary woods were principally that young trees were easier to carry and required no fashioning before use.

Le Roux (1948, p. 446) supposes that it was in these conditions of altered forest that the composite hafting style proved more convenient, since the cutting edge of the blade could be turned to different angles according to a variety of requirements. In evolutionary terms he thought that the axe with the perforated handle, such as has been found in the Swiss Lake Dwellings and northern Europe, was the original one in relation to the adze, though as we have seen previously the composite adze was known in Egypt, Switzerland and Denmark in prehistoric times.

When steel axes came into New Guinea, however, they were found to be desirable not only by people who had no option but to fell large trees in primary forest, but also by inhabitants of secondary forest and grassland, for whom they were useful in many tasks of daily life, like splitting firewood, pointing stakes and clearing undergrowth. In addition the steel axe was easier to haft and the hafting lasted longer.

5. Differentiating between Adzes and Axes

The difference between an axe and an adze is that the cutting edge of an adze is at a right angle to its shaft or helve, whereas the cutting edge of an axe is parallel to its shaft (Fig. 13).

Technically the adze is related to the hand-chopper. The normal chopper is a one-hand tool. The cutter uses his forearm as a kind of helve and stands face-on to the tree he is felling. In *Man the Tool-Maker* Oakley shows a photograph of an Aboriginal of the Pitjendadjara tribe of central Australia cutting a tree with an unflaked piece of stone which had a naturally sharp edge (Oakley, 1952, p. 6, Fig. 1). His stance is the same, facing the stem directly, but he grasps the stone with both hands because of its heavy weight (Fig. 29).

The user of an axe, on the other hand, stands to the side of the cut he produces in the tree, and as he fells it he must tilt his body. Consequently, professional lumbermen using modern steel axes change position from one side to the other of the horizontal cut in the

Fig. 29. An aborigine of the Pitjendadjara tribe, South Australia, cutting a tree with an unflaked piece of stone with a naturally sharp edge. After Mountford (1941). Reproduced with the permission of the Royal Society of South Australia, Adelaide.

tree so as not to overstress their bodies. This difference may be another reason why indigenous people in the most wooded parts of New Guinea preferred stone adzes instead of stone axes when felling the large trees of the rainforest (cf. map Figs 18–19).

How can unhafted blades of adzes and axes be differentiated? Semyonov and his co-workers have been concerned with this problem, experimenting with a variety of tools and studying under the microscope the wear marks produced on them (Semyonov, 1964, 1975). Semyonov thinks that grooves, observed under × 4–10 magnification and running mostly diagonally at an angle of 20–25° to the vertical axis of an unground symmetrical lenticular flint tool, were produced by its use as a hafted axe (Semyonov, 1964, p. 123). He states that the two features,

> the bifacial and diagonal arrangement of the striations are the crucial functional criteria of an axe.... With an adze the traces arise in quite

> a different way. It [i.e. the blade] is fixed to the handle with the [cutting] blade-edge at right-angles to it [i.e. the handle] while the tool's [i.e. the blade's longitudinal] axis is at an acute angle to the handle, not at a right-angle as with an axe. Consequently the wear traces on an adze are sharper on its front face away from the handle, precisely because this side encounters direct resistance from the worked material. In addition the striations lie, not diagonally, but more or less parallel to the adze's axis.... The axe's curved trajectory leaves its marks on its working face as it sinks into the material.*

The implication of what Semyonov is saying is that the cutting edge of the axe hits the trunk with its inner corner first, while with the adze the trunk is struck by the full edge.

Semyonov and his co-workers felled a fir tree, 25 cm in diameter, in 20 minutes without any previous practice using a Neolithic nephrite axe, and he remarks that a "Neolithic axe flaked off the wood not by blows into it but by angle blows longitudinally along the surface. So in many Neolithic axes the front angle is worn away, and the blade looks lop-sided", but "the lop-sidedness occurs not only on axes but also on adzes and even chisels" (Semyonov, 1964, p. 129). This is the feature we have called "skewedness".

Semyonov's statements must be compared with the experience gained from New Guinea and from the experiments in Draved Wood. Jørgensen and Troels-Smith observed that the coarse parallel striations on thin-bladed Neolithic axes were due to the grinding of the blade, whereas working in wood produced a distinctive gloss which obliterated the striations (Jørgensen, 1974, pl. XI,I). This gloss corresponds with the brilliant polish produced on flint sickles during the work of harvesting grain (Curwen, 1927, 1930, 1935; Steensberg, 1943, pp. 5, 141). Some thoughts on polish, through use and associated wear on Australian flake tools, have recently been offered by Kaminga (1977).

When Jørgensen and I inspected the axe blade hafted by Ul at Kuk in 1975 under × 8 magnification, a distinct striation caused by the resharpening of the blade was observed. On the blade from Puyakira'go hafted by Dogore in 1975, the striation from resharpening was mostly at an angle of 20° to its longitudinal axis, parallel to the stratification of the rock from which it had been made. But some

*The English translator does not seem to have fully understood the Russian text. The additions in square brackets are for clarification. In official Russian publications in English the author's name is spelled more correctly Semyonov instead of Semenov as in the title of his English book.

striation was observed in other directions as well, and it was not confined to the cutting edge alone. Because the blade from Kuk had been used only for a short time after it had been resharpened, the gloss of wear was only faintly visible at one place on the cutting edge.

In a study of a large collection of stone blades from the Western Highlands of Papua New Guinea, Lampert (1972, pp. 7, 9) was able to separate them into those symmetrically bevelled at the cutting edge and those asymmetrically bevelled. He was unable, however, to say that the former were hafted as axes, the latter as adzes, because the evidence from museum collections of hafted tools from the region did not lend support to the idea, while the wear marks on the 25 tools on which they were visible (17 symmetrically bevelled and eight asymmetrically bevelled blades) were all of the one type in Semyonov's terms.

The different way in which the cutting edges of an adze and an axe meet the trunk according to Semyonov's experiments may be due to a special technique of using axes in Russia, for in Denmark, as well as in New Guinea, the behaviour of the cutting edge of both adzes and axes seems to be identical. However, Semyonov's observations are intriguing, and it would be interesting to have more details of the size of the microscopic grains of silica in local kinds of wood in order to compare the striations with their causes. Detailed descriptions of the cutting experiments themselves would also be welcome. As Semyonov states, the stone artifacts were used for many purposes, and today the adzes and axes of New Guinea have many roles. The owner normally only possesses one hafted specimen (Hughes, 1977, p. 174), so that the traces observed on them may be the result of different uses. For example, Strathern (1969) relates that a hafted axe could be used in an adze fashion, the worker grasping the blade itself and in effect ignoring the handle.

6. New Guinea Stone Blades Compared with Neolithic Specimens from Europe

Axe and adze blades are usually classified on the basis of their material, shape and measurements. Such systematic descriptions are necessary in order to compare products from one culture to another and from one quarry to another as recently demonstrated by Hughes (1977, p. 135).

However, the complexity and variety in the form of stone tools do not reveal much about the function. As White has said: "I would in fact ... suggest that the majority of stone tool forms were not necessary, in a utilitarian sense, at all" (White, 1977, p. 26). A simple tool-kit can be used in a very sophisticated way in a technically highly developed husbandry as Lerche and I observed in New Guinea in 1968 (Lerche and Steensberg, 1973; cf. Golson, 1977a).

An important factor which is mostly ignored in the classification of adzes and axes is their differing states of completeness. Modern users of cutting tools are very keen on the sharpness of the cutting edges, resharpening these every day if in constant use, and so are the indigenous people of New Guinea. Much time is spent at the grinding stones, and one can guess that our Neolithic forefathers were just as aware of the need for re-grinding their stone adzes and axes as people are today. It is also evident that the edge will be damaged by use, and that heavy damage will have to be ground away, sometimes requiring several days of work. This may be one cause of the existence of asymmetrically curved edges. It is not only the normal daily wear and resharpening which result in the shape of a stone tool as it is found in the earth or stored in a hut in New Guinea.

In the present study we are first and foremost concentrating on the cutting edges, because all the effort of the user would be futile if his tool did not bite effectively into the wood.

The edge of a tool made of flint or chert or volcanic rock is more fragile than that of an iron axe. Stone working blades are therefore rather thick. We have already seen that the two faces may not be symmetrically shaped around the cutting edge, which tends to prevent the blade sticking in the tree during felling. There is another characteristic of some blades which deserves mention. Aufenanger and Höltker, who described the culture of the Gende people before the Second World War, noticed that some blades were furnished with a facet on one face about 1 cm broad along the cutting edge (Aufenanger and Höltker, 1940, p. 35). A similar facet was observed by Semyonov on a nephrite adze from a Neolithic burial on the river Angara in Siberia (Semyonov, 1964, Fig. 61, 3), and it corresponds with what Lampert and Hughes call a "chin" (Fig. 30) (Lampert, 1972; Hughes, 1977, Fig. 28).

The cutting edge is facetted on the face which would be turned to the right of a right-handed man when cutting a tree, and the reason for this is the same as that for facetting a modern plane blade: the edge can be resharpened without making it too thin and fragile. This is an age-old principle. In the late Pleistocene, people on Kangaroo Island, then part of the South Australian mainland, flaked water-worn

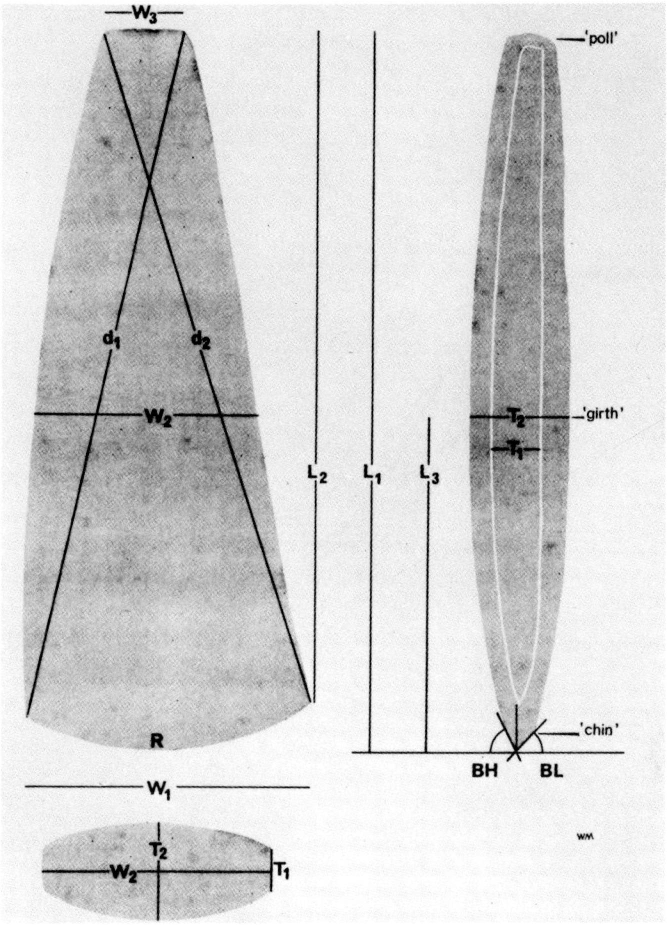

Fig. 30. Mount Hagen blade with a facet or "chin" at one side of the cutting edge. After Lampert (1972). Reproduced with permission.

pebbles unifacially to produce an edged tool or chopper. These must have been used with their naturally smooth side against the tree-trunk and the chipped side could then easily be resharpened when it became blunt. The pebble chopper shown in Fig. 31 has a kind of "chin" or facetting (Lampert, 1977).

It is useful to compare flake-axes from the Mesolithic of Denmark with the stone blades with a facet along one side of the cutting edge. Troels-Smith has classified a number of the former and drawn the piece shown in Fig. 32 (Troels-Smith, 1937). Its flat face obviously

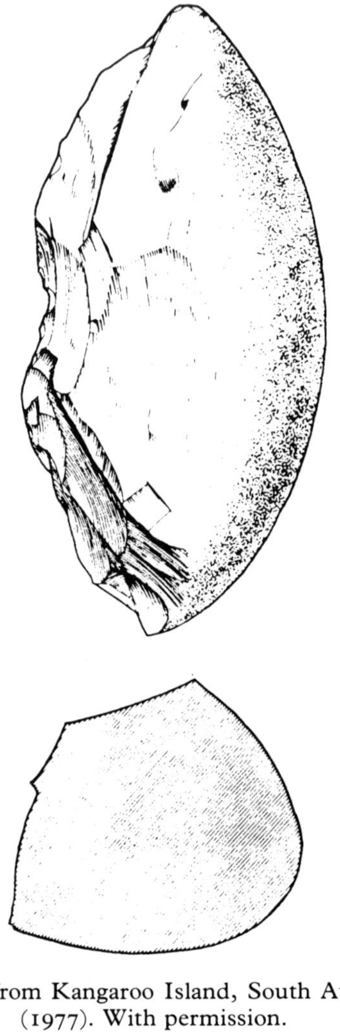

Fig. 31. Pebble chopper from Kangaroo Island, South Australia. After Lampert (1977). With permission.

faced the trunk, whereas the curved surface of the other side corresponds with the edge-facet or "chin" of the axe from Mbukl as well as that of the chopper from Kangaroo Island.

The stone blade of the axe hafted by Ul at the Kuk Tea Research Station in 1975 was badly damaged and had to be resharpened. In so doing Ul produced a distinct facet on the cutting edge, and in Fig. 33, upper left, the axe blade is viewed from this side. Below this is a side view from above and the facet is seen on the bottom right on

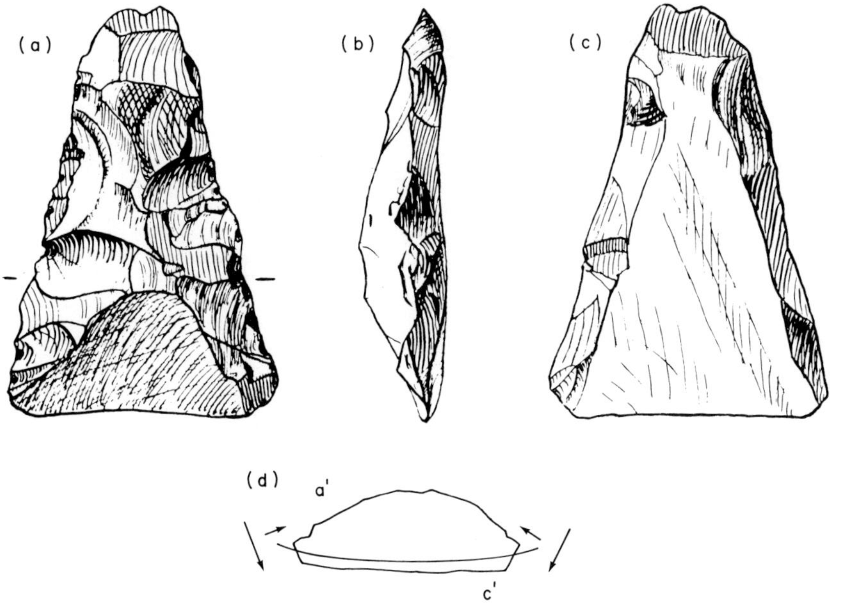

Fig. 32. Symmetrical flake axe, Late Mesolithic, Denmark. After Troels-Smith (1937). With permission.

the drawing. In the view on to the cutting edge, which reveals more of the characteristic of such an implement than a simple median cross-section, the facetting is shown on the left of the edge. It is also obvious from this view that the edge is not straight but slightly S-shaped.

The facetting of stone blades can also be observed on some European thin-bladed axes from the early Neolithic. A Danish flint blade found in Jutland is shown in Fig. 33, lower left. This has a similar facet on the side, which during use would have been away from the trunk, though it is not so distinct. The facetting may be due to resharpening—intentional or by chance? The other specimens from the Danish Neolithic shown in Figs 33 and 34 are also slightly asymmetrical viewed from the narrow side, looking almost like the mouth of a pike or a cod. The side away from the trunk is characterized by a more abruptly curved profile to the cutting edge.

Which face of a blade faced the trunk and which faced away can therefore simply be observed from the fashioning close to the cutting edge, especially when viewed from the narrow side. Moreover, the face against the trunk may be rounded, especially if the cutting edge is concave as it often is on an adze. In Fig. 35 the Puya-kira'go

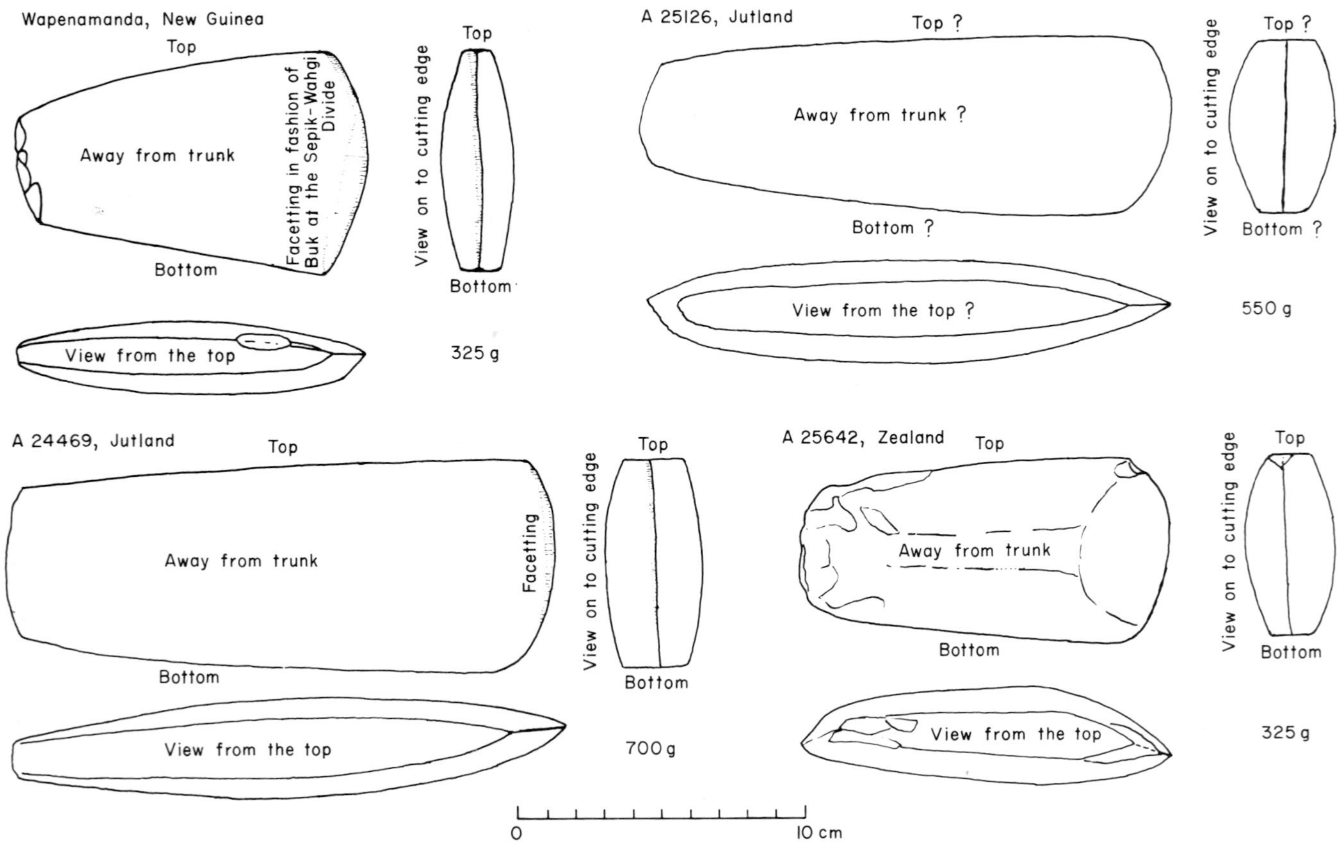

Fig. 33. The axe blade from Wapenemanda (Fig. 12) sharpened in the Hagen fashion with a facet or "chin" (top left) compared with three Neolithic flint axes from Denmark. Drawn by the author.

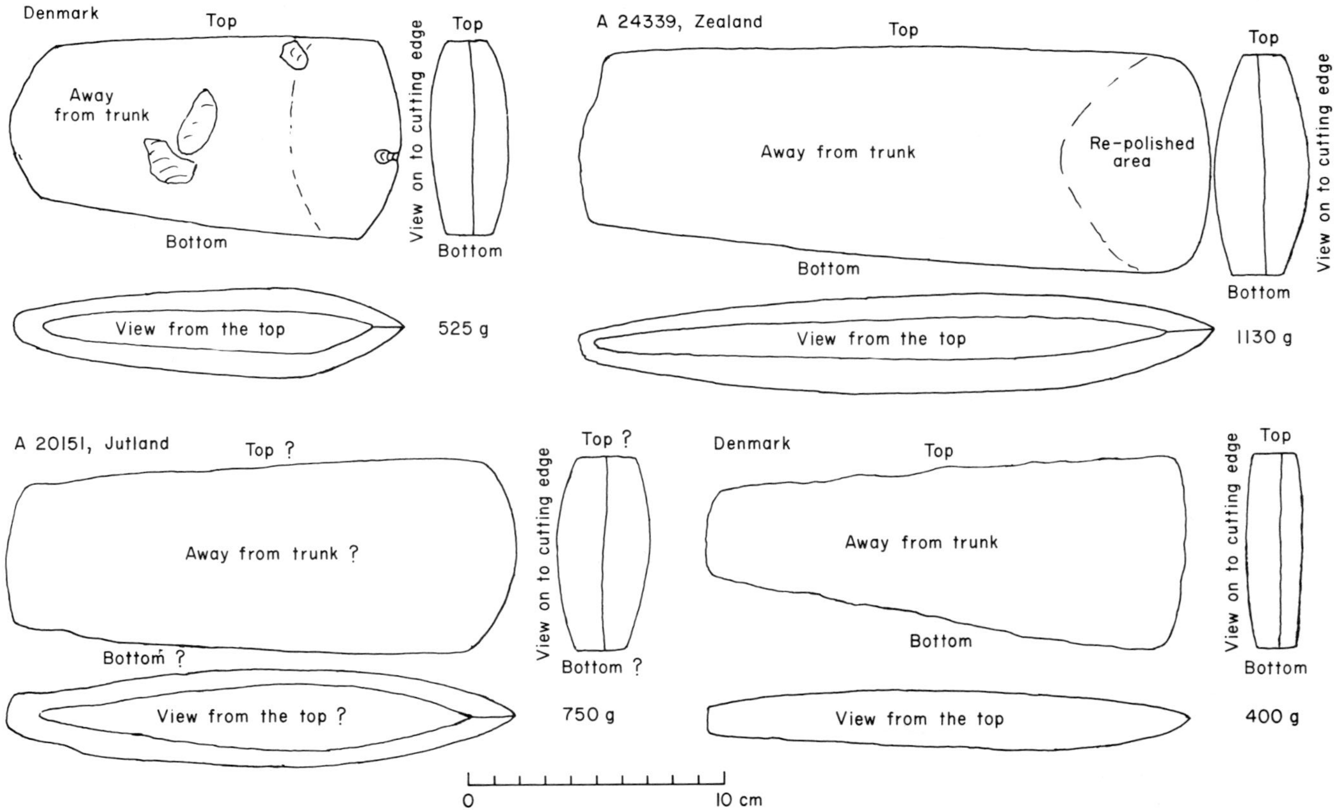

Fig. 34. Danish Neolithic axe blades. Drawn by the author.

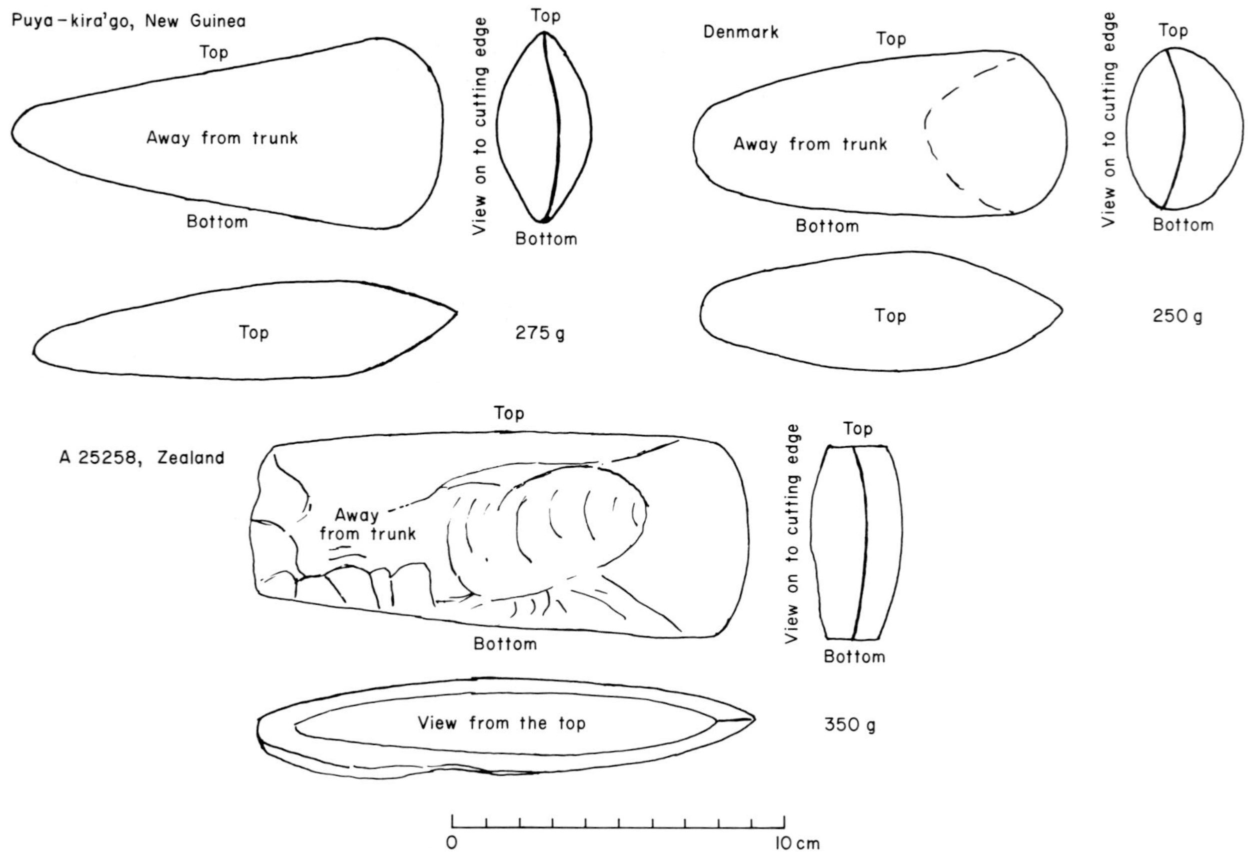

Fig. 35. The axe-adze blade of the Puya-kira'go tool (Fig. 11) compared with two Neolithic adzes from Denmark. Drawn by the author.

blade is compared with two Danish Neolithic specimens, the one to the right made of greenstone, the other below made of flint. The greenstone one may possibly have been hafted as an adze because of its concave cutting edge, though it could also have been set at an oblique angle to the shaft similar to the specimen from Puya-kira'go. The lower specimen, No. A 25258 from Zealand, in the Danish National Museum, could hardly have been hafted in a perforation of the helve like the axe shown in Fig. 13 because its 5·5 cm wide butt or neck end would have required a very clumsy shaft. But because the cutting edge, viewed end-on, is concave like the other two specimens in Fig. 35, it could have been hafted as an adze to an angled haft in a manner similar to the implements of the New Guinea highlands and to some Swiss hafted specimens found in the Lake Dwellings.

According to the arguments advanced above, when a right-handed man used a hafted blade, the flat side of the blade which faced the trunk would be on his left, and the more rounded side which faced away from the trunk would be to his right. When I was studying hafted Neolithic axes from the Swiss Lake Dwellings in the Historisches Museum in Berne, I noted that a few tools, if used right-handedly, would have had the flat side of the blade away from the trunk. These tools can only have belonged to left-handed people.

The weight and momentum of a stone blade is another property which is worth considering because the man who is hafting it must have an idea of how the tool will suit him best. Differences in age, experience and strength are of course decisive factors in making a choice, though a full adaptation to the needs of the user cannot be expected. A good blade will not be thrown away if damaged. It will be resharpened and its hafting will be renewed many times, so that it can be adapted to the blade successively, each time restoring the convenient balance of the tool. Therefore it is not just a question of producing a tool which has a heavy momentum for felling trees. It is just as important that it is properly hafted, so that the lumberman can exercise all his strength while cutting, without being worried about breaking his tool. In this respect there proved to be a great difference between the axe hafted in Kuk and the axe-adze hafted in Puya-kira'go, the last one being exceptionally strong.

Table I gives a comparison of the weights and blade lengths of New Guinea tools with those of the Danish Neolithic (Figs 33–35), together with two specimens used in the Draved experiment.

The variety of size and weight of blades is considerable as one would expect from the above discussion about resharpening and differences in age and strength of the users. But the maximum weight of specimens from what is now Irian Jaya, according to Le Roux (1948, pp. 415, 432), is not much less than the heaviest specimen illustrated in Fig. 33 from the Danish Neolithic. According to Jørgensen, short, heavy blades proved to be most effective in the Draved experiment (Jørgensen, 1974). We can see from what has been said that there is hardly any difference between the stone axe blades of present day New Guinea and Neolithic Denmark, apart from the materials which in Denmark are mostly flint and in New Guinea volcanic rocks.

Table I

	Weight of blade	Weight of helve	Weight total	Length of blade
Axes from Irian Jaya according to Le Roux	(g)	(g)	(g)	(cm)
Max.	850	900	1750	20–25
Average			1250	
Axe-adze from Puya-kira'go	275	975	1250	13
Axe hafted in Kuk	325	925	1250	12
Axe hafted in Wapenamanda			*c.* 1000	*c.* 14
The same axe in a dry condition			850	
Denmark, Neolithic				
Original axe from Sigerslev (helve restored from pieces)	915	No information		21
A 24469	700	No information		19
A 25126	550	No information		18
A 25642	325	No information		13
A 25258 (adze)	350	No information		15
No number	525	No information		15
A 20151	750	No information		20
A 24339	1130	No information		25
No number	400	No information		19
Denmark, Neolithic				
Axes hafted for Draved experiments, constantly used without damage but whetted frequently				
II A 1503, copy of Sigerslev hafting	606	516	1122	
M_I-b, copy of Sigerslev hafting	*c.*638	*c.*540	*c.*1178	

7. Clearing Grassland

Although the techniques of making swiddens have been described repeatedly by geographers and anthropologists, nobody has explained in detail how the tough sod of grassland cover was cut and cleared away before the introduction of bush knives, axes and spades of steel. Yet it is generally accepted that extensive areas had been transformed from forest to grass as a result of the growing population and expanding horticulture in the New Guinea highland areas, before the arrival of steel tools (cf. Golson, 1977b, pp. 602–604). In the Hagen district a heavy digging stick, up to 3 m long and 10–15 cm in diameter, was formerly used to slash the undergrowth as well as for digging up stumps of trees and cane grass (of various types, called pitpit in Pidgin) and for digging the fenceline (Powell *et al.*, 1975, p. 13). For clearing areas of short grass a smaller paddle-shaped spade was used to remove the surface turf.

Watson (1967, p. 97) mentions that in Kainantu, in the Eastern Highlands Province, he was told that before the Europeans introduced steel cutlasses people had a bamboo implement specifically to cut grass. No original tools of this kind could be obtained but two men made new ones for him from longitudinal sections of bamboo of perhaps 13–15 cm in diameter.

> Both edges were sharpened except for a short distance near one end. This formed a handle, slightly narrowed and capped by a knob. One such knife was 30$\frac{1}{4}$ in. [76·84 cm] long, its blade a maximum 2$\frac{1}{4}$ in. [5·72 cm] broad, the other 32$\frac{3}{4}$ in. [83·19 cm] and 2$\frac{3}{16}$ in. [5·56 cm] respectively [Fig. 36]. As far as could be judged, both specimens appeared authentic to those who remembered using the tool in the past. The knife was used in cutting the grass to dry for firing a new garden plot and possibly also for clearing a site covered in grass for locating a new house or settlement ... I do not know how widespread ... the bamboo grass knife ... may be.... The grass knife may have been

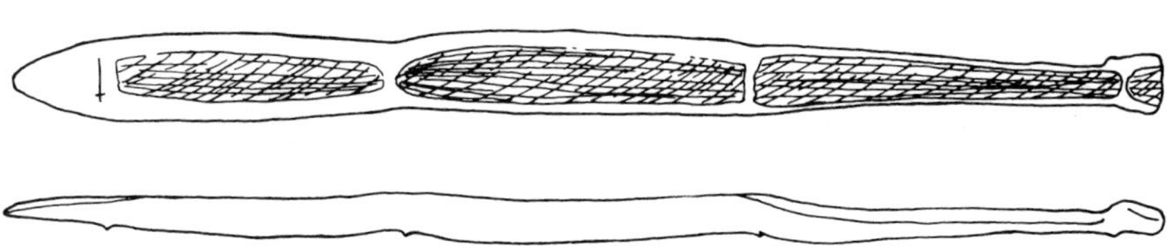

Fig. 36. Bamboo grass knife from Kainantu, Eastern Highlands Province. Side and edge views. After Watson (1967, p. 97, Fig. 2). Reproduced with permission.

unknown or of less importance in areas of bush gardening as opposed to grassland. A friend from the coast of Papua said he readily recognized the grass knife. (Watson, 1967, p. 97)

A similar bamboo knife seems to have been used by the Hiowe, because, according to Townsend, the steel bush knife was called *ipari mene'e* which means "foreign bamboo" (Townsend, 1969, p. 201). Neuhauss mentions that in the drier part of the Huon peninsula grass was cut with bamboo knives or with swords (Neuhauss, 1911, Vol. I, p. 283).

One day during my stay at the Kuk Tea Research Station in 1975, Philip Hughes discovered two wooden tools from a store where finds made during drainage of the local swamp had been occasionally put away. These were shown to the men from Mbukl, who told us that they were predecessors of the steel bush knife used for grass cutting.

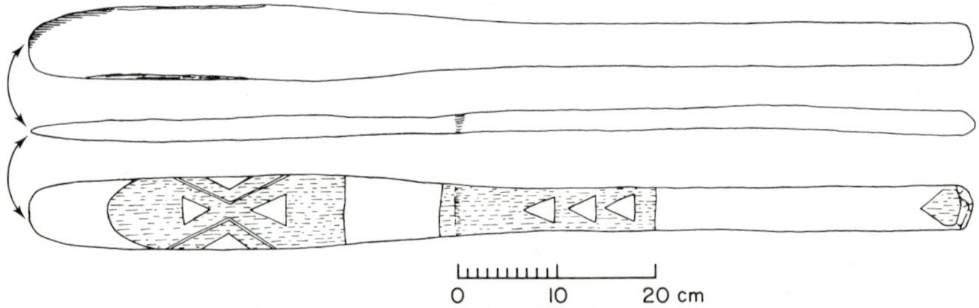

Fig. 37. A grass cutting knife called *yakla* after the palm from whose wood these implements were made in the Mount Hagen area before the introduction of steel bush knives. This example was made for the author in 1975 and drawn by him.

We decided to have such a bush knife made for demonstration, so one day we went with Ul, El and Ru to the forest of the Wahgi-Sepik Divide and on a slope felled a *yakla* palm as previously described. The men said that this was better than casuarina wood for making bush knives and paddle-shaped spades, and from it they manufactured a couple of *yakla* knives. One of these knives, shown in Fig. 37, was 95 cm long, maximum width 7·3 cm and thickness 2·5 cm. It was made from the outer wood of the palm, and on its curved surface Ul produced a geometric decoration by cutting away parts of the bark. The inside of the blade was slightly hollow.

When grass is cut with bush knives, one man slashes at the roots while one or two others pull the loosened grass cover away like a carpet, using wooden hooks about 150 cm long called *mak* in the local language. Hooks are easily made from a forked branch of the casuarina or *nde kaepa* tree and they are never found in the swamp, probably because, being easily made, they were not kept in water or mud like paddle spades. Such hooks are frequently found from the Neolithic in north European bogs and the Swiss Lake Dwellings, where they are erroneously said to be prototypes for the ard-plough. Probably they were used as rakes to gather leaf-fodder and similar material.

In our first demonstration with *yakla* knives in the drained swamp basin at Kuk, four men used the knives and six pulled the cleared grass away. In Fig. 38 Ul has just completed his cutting movement, while Arit is ready with his hook to clear the vegetation away—in this case 3–4 m high and 2 cm thick cane grass. It was not long before the men began to show signs of exhaustion. They had started at 9:35 in the morning and by 10:45 they were all tired. At 11:35 they had cleared little more than 100 m^2, though it must be noted that one of the younger men had a bad cold. However, the short swamp grass proved much easier to deal with (Fig. 39).

At midday, Ru announced that the work was wrongly organized. As far as he remembered, they normally cut the heavy cane grass near the roots with stone axes after the stems had been bent over with a short heavy stick underneath as a chopping block. When we tried it we found that when a great mass of stems had been bent over, it was easy to cut the stumps just above the roots and pull them away with the hooks. Ru demonstrates the procedure in Fig. 40, using the stone axe made by Ul. Jørgensen tells me that in the Draved experiment they learnt that slender stems, which were bent rather than cut by a blow from a stone axe, had to be dealt with in the same way that Ru employed for the cane grass at Kuk, that is using a heavy stick as a chopping block.

Then Ru showed how the short swamp grass cover could be opene
up using the stone axe and stick in the same way (Fig. 41). With hi
left hand he put the stick between his legs from behind and cut t
grass with the axe in his right hand. After a line had been open
up in this way it was easier to cut the roots with the *yakla* knif
the grass carpet was pulled back with hooks. The cutting of both c
and swamp grass was filmed.

The *yakla* knives had to be sharpened at least twice a day be
the sharp edges became blunt during use. As shown in Fig. 3
edge on the left of the user would be worn most when the ba
faced upwards. Sometimes the men would turn the tool so t
slightly hollow inside faced upwards, mostly because the sha
(at the top in Fig. 37) had become dull. However, they pre

Fig. 38. In a demonstration in 1975 at Kuk Tea Research Station, Ul from Mbukl used a yakla knife on the thick cane grass cover, while Arit cleared the cut vegetation away with a one-pronged rake or hook.

Fig. 39. Two men cut at the roots of the short swamp grass with yakla knives, while two others pulled it back with hook-shaped rakes.

Fig. 40. Ru cuts the heavy cane grass near the roots with a stone axe using a heavy wooden stick as a chopping block.

Fig. 41. The short swamp grass cover was also opened up using a stone axe and stick. Ru is working while Ul stands in front with his yakla knife on his shoulder.

have the bark side upwards, and therefore one edge was worn more than the other before both edges were resharpened. This process was carried out using a steel bush knife. The *yakla* wood is tough but its fibres are rather coarse and loose, though being siliceous it can take a fairly sharp edge.

Le Roux mentions how the people in the Swart valley north of the Baliem river used to pull up the alang-alang grass (*Imperata cylindrica*) with their hands, spread it out on the surface to dry and afterwards burn it (Le Roux, 1948, p. 284).

In a recent publication Golson (1977a, p. 155) summarizes the literature on the use of stone knives as machetes for different purposes. He includes a reference to Bulmer (1964, pp. 252–253) whose contact from east central Chimbu said that stone blades with wide cutting edges and sides converging somewhat to the poll were used within living memory for cutting cane grass. On his journey to central Irian Jaya in 1921–1922, Wirz (1924, pp. 108–109) observed that the indigenous peoples made stone knives of different shapes and sizes. Some were ground on both edges, others were ground only along one edge. Those ground along both edges were shaped like daggers, while those with only one edge ground were often sickle-shaped. All of them were used for dressing timber, manufacturing arrows and other weapons (though never as weapons themselves), and peeling tubers. Le Roux (1948, pp. 423–424, 450–451) records their lengths as ranging from 3 to 32 cm, the breadths from 2·5 to 7 cm and the thickness from 4 to 12 cm [sic!]. He also states that the women used them for cutting vegetables and harvesting crops from the field and from trees. Men used them for cutting bamboo, rattan and withes, which they stripped of twigs, leaves and thorns. Neither Wirz nor Le Roux seem to have seen stone knives used for clearing cane grass, which is apt to make one wary of generalizing too much from a single piece of information.

Of course, not every area presented the necessity of grass clearing. According to Clarke, in the Ndwimba basin of the Simbai river in the Bismarck Mountains, grassland is the rarest type of regrowth, covering less than 5% of the total area occupied by secondary vegetation. "It occurs only on the driest sites on the top of spurs, and its small presence there does not point to the permanent conversion of forest to grass that has happened in much of New Guinea." In this valley the grassland seems to be only a transient stage in the regression to forest. *Imperata cylindrica* grows in the driest sites. It is difficult to clear, and afterwards it has to be weeded away because it sprouts persistently from rhizomes (Clarke, 1971, p. 65).

In areas where a marked dry season exists, grasslands are burned, though high pitpit is difficult to clear in this way. The roots can be removed by digging with heavy digging sticks. Formerly, the men often used to work in teams with long digging sticks as illustrated by Heider for the Dugum Dani in the Baliem valley (Heider, 1970, p. 41, pl. 7). At this point it is worth noting that teamwork in the use of wooden spades and digging sticks was spontaneously invented in many parts of the world, for example, Gade and Rios record the same practice in Peru (Gade and Rios, 1972, p. 12, Fig. 9), and Damm (1955, p. 103, Figs 3 and 4) records it in Indonesia.

In the Atlantic region of Neolithic Europe open grassland hardly existed outside some marshy coasts and shores of lakes. The European meadows were an artificial creation during the Iron Age, when the introduction of iron smelting made it possible to produce tools suitable for the production of hay. Longer and narrower scythes were developed, adapted to the successively opened up low areas around rivers and lakes, until in the late Iron Age long-shafted scythes were used regularly (Steensberg, 1943). Some archaeologists (e.g. Jarman and Bay-Petersen, 1976, p. 181) have presumed that Neolithic people harvested hay, but the conditions for cutting grass hardly ever existed. Instead, cattle were fed on twigs and leaves while stalled, or kept inside fences as is practised in the middle of Switzerland even today (Troels-Smith, 1955, p. 53). The fodder was probably cut with small knives of flint.

8. Slashing and Burning

As stated in the introduction, the aim of this study is not to give a comprehensive survey of shifting cultivation in New Guinea in the sense of Conklin's general outline (Conklin, 1963), or of an earlier survey by myself (Steensberg, 1955). The intention is rather to compare some present-day New Guinean methods of slash and burn husbandry with the evidence of similar practices from the European past.

Clearing and swiddening of uncultivated areas were known in antiquity in most parts of Europe and Asia west of India. According to Arrianos, the Roman governor of Cappadocia, the Oritic tribe near the outlet of the Indus, in what is now Baluchistan, used to fell the trees with "sharp stones" because iron tools had not yet been introduced (Arrianos, Indice, Chapter 24). Under such conditions in antiquity, as nowadays, people had to take advantage of whatever practical means the environment could offer. They cut the trees at the most suitable height above the ground, using scaffolds if necessary, and they preferred gently sloping hillsides where the so-called "windrow-felling" could be practised.

It is easier to fell trees down a slope, and to cultivate from the lower edge of a sloping field or garden upwards rather than downwards. Therefore, the earliest forest clearances were normally placed on the slopes of hills with easy access to rivers and lakes, as can be learned from areas where swidden agriculture exists today. Sometimes a number of trunks were cut half through, so that they would all fall at once, rather like an avalanche, when one or two big trees were cut down at the top of the swidden. This method of "windrow felling" produced a triangular-shaped clearance area, the base of the triangle being at the bottom of the hill. In the interlaced rainforest of the tropics in particular this method resulted in a considerable saving of labour (Steensberg, 1955, pp. 70, 72; cf. Le Roux, 1948, p. 449).

Koch-Grünberg (1910, p. 202) describes the practice of this technique in Amazonia. When a young man intended to marry, he would clear a considerable area of wood near the river for a garden, beyond the reach of the annual flood. Initially he would half cut the smaller trees, then fell the larger ones, which in their fall would bring everything down, partly because of creepers which connected all the trees together. Another German traveller, Schmidt (1905, p. 102), observed the practice in the upper reaches of the Xingú and Tapajóz, tributaries of the Amazon. Hose (1926, p. 117) described a similar procedure in Sarawak, and Takács (1977) relates that windrow-felling was also practised by Hungarian assarters in the nineteenth century.

If there are navigable rivers in a primeval forest these will always be the easiest means of transportation. And for clearing and cultivation, an undulating landscape will be preferred to a flat one. This must also have been the case in Europe in the early Neolithic. During my previous studies on the distribution of the burial mounds and megaliths in Denmark, I became aware of the significance of these two factors for the first agriculturists in the country (Steensberg, 1937, 1943, 1950). Plains such as the southern part of the island of Lolland and the bay of Køge in eastern Zealand did not attract the settlers. The megalith builders preferred to spread into central Zealand along the river Suså, and settled here around some lakes. But they made their clearances particularly in the forest around the peninsulas and inlets of northern Zealand as well as along the straits of Jutland and Bohuslen on the east coast of the Kattegat.

The Danish archaeologist Müller suggested that where the burial mounds of Jutland were arranged in long narrow zones there must have been main roads between settlements (Müller, 1904). However, because burials were made where people lived, it seems more likely that these zones of burial mounds simply indicated settlement where environmental conditions were particularly favourable. They proved to be huddled together particularly along the limit of the last glaciation, where hilly slopes face sandy plains produced by glacial outwash. Apparently, prehistoric farmers had chosen an adequate niche here

Fig. 42. A new clearing, recently fired, in the secondary forest near Yainya Creek, west of Lake Kopiago. In front is a women's house and behind a tall *Araucaria*.

at the boundary between the dense climax forest and the light scrub on the glacial sands. This is an example of how knowledge of factors of location and the techniques of recent stone age people may contribute to a better understanding of the European past.

In rocky places, such as the karst landscape around Puya-kira'go, swiddens cannot always be placed on gently sloping hill sides. I was not able to observe the burning itself because the season was too wet. In fact there is no really dry season at Puya-kira'go; the driest months, relatively speaking, are December, January and February, and it is normally during these months that any burning takes place. However, on my way from Kopiago I saw a very new clearing situated on a slope east of Yainya creek (Fig. 42). In the foreground of Fig. 42, a women's house and flourishing sweet potato beds can be seen, and on the slope beyond a little valley there is a small swidden which had recently been cleared and burnt in spite of the wet season. A tall *Araucaria* rises above the secondary forest, which had not yet grown up to its full height after its previous felling. In the distance are the mountains behind the Tumbudu river, a tributary of the Strickland.

Another clearing in the same region had not yet been burnt (Fig. 43). Some slender trees remained standing, pollarded so that their branches would not shade the ground. Others had been cut down and lay scattered between ferns, branches and grass which would soon be cleared too. In the lower part of the picture the main twisting trail connecting Puya-kira'go with the Kopiago Patrol Post can be seen. To the right some men rest under a tree which has not been pollarded. Here and there stumps of felled trees stick up. When the area has been burnt some day, the spot will be fenced against feral pigs.

In the Ndwimba valley on the north-eastern slope of the Bismarck Mountains, Clarke noted how the Angoiang men felled heavy trees using scaffolds (Fig. 44). The two trees in the centre were the only ones surviving from an earlier period than the surrounding secondary growth and after felling them, the men slashed the undergrowth and pulled up the seedlings, spreading them on the ground to dry. When the material was dry enough to burn all the litter was piled up and ignited. When a pile was burning well, the burning material and ash were spread over the surrounding ground with a digging stick. But much undecomposed litter would remain unburned, and only about 30° of the surface would be covered with ash (Clarke, 1971, pp. 144, 149).

This technique mirrors the practice in Europe in recent times. In Finland as well as in Sweden it was common to cut off all the branches

Fig. 43. A new clearing, not yet fired, in the same region as that in Fig. 42. The unfelled trees are pollarded so as not to produce shadow over the new garden.

Fig. 44. This man had erected a scaffold to fell a buttressed tree with a steel axe. Ndwimba basin, south-eastern slopes of the Bismarck Range. Photograph by William C. Clarke, reproduced with his permission.

and twigs of the trees in order that the fire could be kept as close to the surface as possible. When the fire had run over a field, as slowly as possible and always against the wind, unburned pieces of wood were collected into small heaps and thoroughly burned. Afterwards the hot ashes were spread on the surface as evenly as possible. Though the Scandinavian countries and Finland are mainly situated in the region of coniferous forests, the regularly swiddened areas around villages and in the more densely populated areas were largely covered with birch because of repeated slashing and burning. In Savolaks, Finland, the peasants started the first burning in May–June after the wood had been cut the previous autumn before the leaf fall. The half-burned pieces were then arranged in rows so that the soil could soak up the lye of the ashes until the following summer. Then the rows of half-burned material were ignited one after the other. The burning material was pulled over the ground so that the soil would be equally burned and none of it "dead-burnt", and the effect of burning would not penetrate more than 5–7 cm in depth (Steensberg, 1955, p. 79).

The differences in burning techniques depend of course primarily on the different climatic and vegetation conditions, but the difference between the climax forest and the secondary forest also influences the techniques of felling and burning. One of Clarke's Bomagai-Angoiang informants said that with steel axes they felled large trees that they formerly would only have pollarded with their stone axes (Clarke, 1971, p. 189). In those times it must have been hard work to extend the swiddened and cultivated areas into the climax forest, though Le Roux describes how this was practised in the Nògòlò river area in Irian Jaya (Le Roux, 1948, pp. 448–450). Here large trees with buttressed roots were ringbarked with the stone adze, and a pointed stick was used to loosen the bark-strip in the same way as the pointed axe-helve at Kuk, mentioned above. When the ringbarked trees had lost most of their leaves, they were felled by fire.

In Puya-kira'go, Irari demonstrated how to kill a tree by putting fire to its roots. He took a fire-brand from the men's house and carried it some 100 m into the wood where he made a bonfire around the roots of a tree (*mbuku? puku?*), about 25 cm in diameter, which had been ringbarked two years previously (Fig. 45). It had still not lost all its leaves and if left to itself it would have died in two to three years' time. Firing, however, accelerates this and the tree would have lost its last leaves within three or four weeks. Giant trees could be felled exclusively by fire as Wirz observed on his travels in Irian Jaya in 1921–1922 (Wirz, 1924, p. 88). Ringbarking is not an essential pre-

Fig. 45. This tree at Puya-kira'go had been ringbarked two years previously and when the photograph was taken in 1975, it was being killed by fire.

requisite to firing, but its practice has a useful purpose. When clearing the vegetation of an area for cultivation, some large trees are always left as a future store of firewood (see Chapter 5). If they have been ringbarked they will not take nutrients from the soil.

9. The effect of Slashing and Burning on Soil Conditions

From olden times it has been assumed that the ashes produced from the burning of vegetation, either the organic litter of woodland or the sods of a heathland cover, are active fertilizers.

In *La Grande Encyclopédie* (Diderot and d'Alembert, 1751–1772) the terms *brûler ou écobuer les terres* were explained in this way: "When soils, having been resting for years, are cultivated again they should be burned into ashes because the fire will disintegrate the earth and the ashes of leaves and roots will make it fertile." But as recently as 1956, de Schlippe, in describing the shifting cultivation of the Zande people in south-west Sudan, stated that agricultural research had not yet been able to point conclusively to what alterations in physical structure or chemical composition of the soil the loss of fertility, during long time cultivation, could be attributed, and consequently what the effect of swiddening as a means of restoring it could be (de Schlippe, 1956, p. 120).

It was not until Nye and Greenland (1960) published the results of their research in the tropical environment of Ghana that exact figures of the chemical effect of swiddening became available. They proved that this effect is essentially confined to the upper 2 cm of the soil. The carbon, nitrogen and sulphur in the fallow and litter are lost in the burning, while the amount of humus in soil itself is preserved. An important point, particularly concerning acid soils, is that the alkaline ash raises the pH and the cation exchange capacity. In addition, the fire partly sterilizes the upper layer of the soil and destroys most of the weeds which would compete with the plants to be cultivated.

The effect of destroying weeds by burning is more important in tropical rainforests, than it is in the coniferous woods of temperate climates, because of the rapid growth of the vegetation under hot and humid conditions. In the coniferous woods of Scandinavia and Finland a similar effect can be obtained by hoeing, i.e. mixing the uppermost soil layer with deeper ones and getting rid of the roots of the natural vegetation. In fact it is not necessary to burn the wood at all in these forests because of a special "assart effect", which can be obtained by merely cutting down all the trees in a particular area (Romell, 1957).

Although the Finnish assarters knew of this effect 200 years ago (Krok, 1765) it seems to have been a neglected subject until the Swedish professor Romell (1957) drew attention to it after the Second World War. The Finnish peasants had noticed that it was not the ashes that fed the crops so well, rather it was something produced by the acidifying or "souring" of the soil by newly cut roots or trees, a kind of composting. Thanks to this assart effect even the supposed inactive tough "mor" (i.e. sour humus) of old, poorly growing northern spruce woods quickly releases ammonia in great quantity once spruce trees growing on the "mor" have been felled. Samples of such "mor" stored in a laboratory had up to 30% of their total nitrogen available after a year and a half of storage. In comparison, the best Danish beech "mull" (humus) would have less than one-half of this.

Romell (1967, p. 367) believes that this assart effect has a good deal in common with composting and green manuring, and according to him a good "mull" may still exist under a middle-aged forest cover. As the stand ages, the roots may become matted and concentrate at the surface of the soil. This may go so far as to change the type of humus layer from "mull" to "mor", while a light "mor" may become more and more matted by roots and fungal growths, whilst the soil below it hardens and a leached layer develops. This is a fact already observed in Danish beech woods by Müller (1887, 1889). However, Danish foresters have practically done away with the beech "mor" by replacing old stands with young growth and thinning those still remaining.

In Draved Wood, where the experiments in slash-and-burn cultivation referred to previously took place, thick layers of "mor" are still preserved, probably because it has never been properly managed. According to botanical investigations decomposition is much slower in "mor" than in "mull", even if it is exposed to the effect of the air after the tree cover has been felled. Iversen found a yellow layer of so-called "fossil mor" below the upper "mor" layer at Draved. When the "mor" of an oak forest reaches a certain thickness, it is rather resistant to decomposition especially on low ground with a high water table. This may be one reason why the assart effect was not observed in the unburned plots slashed in Draved Wood.

The Draved experiment was carried out at a time when soil science was not as well developed as now, and we would plan the experiment differently if we were able to repeat it today. However, the report published by myself (Steensberg, 1979, Tables 4–6, 10) contains some figures concerning the chemical condition of soils before and after burning, which are relevant to the questions discussed here.

Clarke (1971, p. 68) relates from the Ndwimba basin in the Bismarck area of New Guinea that the median pH of the top 15 cm of soil was 5·5. The most acid soils were found under grassland, where aeration is poor, and under primary forest. The difference between burned and unburned soils was only 0·5. Apparently his samples were mixed from the upper 15 cm of the soil. The Draved analyses proved that samples should be taken separately from the upper 2 cm from the next 2 cm, from the following 4 cm and finally from the lower 7 cm, between 8 and 15 cm depth. Moreover, the pH may differ within an area of less than 0·5 m, because animals often urinate or defecate at a certain spot; or the decomposed corpses of a bird or an animal may raise the pH at one particular place. At least the Draved sampling proved that the pH values differed considerably from place to place within the total slashed area of about 0·5 ha land; and they fluctuated from year to year within the same plot of unburnt soil. Moreover, this oscillation did not follow the same rhythm from one plot to another.

In the burned areas the pH was generally raised as a consequence of the combustion, especially within the upper 2 cm of topsoil. But within three years, the differences had been virtually wiped out. Before burning, most pH values in the topsoil lay between 3·5 and 4·5; after burning, they normally rose to between 5 and 7, and in one particular sample even up to 8·2. However, the pH values in the subsoil were often as high as 5 to 6, so the effect of mixing this soil with the top layer would in this respect not have been much different from that achieved by combustion. However, the existence of matted roots and stones prevented such a procedure in the Neolithic.

The inorganic phosphate level in the Draved soil was measured by the phosphorus index, which also increased remarkably as a consequence of the burning. The potassium index in the upper 2 cm of the soil rose too, but the released potassium content was leached out within two or three years. Regarding the nitrogen content we had the same experience at Draved as Nye and Greenland in Ghana, namely that the small levels of between 1·24% and 1·41% were reduced to zero by the combustion.

In future experiments and analyses of swiddened areas it will be of vital importance to fix the measuring frame with greater accuracy than at Draved, as well as to sample from the upper 2 cm downwards, as described above. Moreover, the grids or section-lines should be arranged so as to cover any possible variation in the plant cover. Also the amount of charcoal and ashes covering the sampling plots should be calculated. However, the analysis of soils is a difficult task and generalizations should be regarded with caution.

Although the climate and vegetation in the rainforest of New Guinea is very different from that in the deciduous forests of Denmark, the superficial effects of slashing and burning would be very similar. This is clearly seen by comparing the two photographs (Figs 46 and 47). The first of them was taken between Mount Hagen airport and Kindeng near the Wahgi river. This swidden was divided into individual plots with stakes and small trenches, and planted with beans

Fig. 46. A swidden near Kindeng in the Mount Hagen area, 1968. Along the road in the background logs are piled up as a protective fence. Charred tree-stumps can be seen in the already planted garden.

Fig. 47. An experimental swidden made in Draved Wood in 1953. The author is making holes for the seed. In the background are charred tree-stumps similar to those in Fig. 46.

and maize, and cooking pits showed that people had had their meals there when cultivating the burned area. Only some dead trees and a number of burnt stumps from trees felled with steel axes were now visible. This is very similar to the other photograph (Fig. 47), from the experiment in Draved Wood, South Jutland in the spring of 1953. Here I am shown making holes for the seed in the charcoal-covered ground, and behind me are several burnt stumps of trees felled with axes of flint before the swidden was fired. They are of approximately the same height as the stumps shown in Fig. 46.

2 The Techniques of Gardening

The monotonous shrill of cicadas and the very high two-note piping of another insect decrease before dawn. The sun rises and the mist will soon disappear from the slopes between Puya-kira'go and Irari Hipuya's gardens far in the distance. The contours of the remote mountains emerge with the ascending sun. They appear greenish dark-brown, covered with rainforest where they hem the valley of the Strickland river. Its turbulent waters turn lazy and sluggish where they join the River Fly with its meanders and ox-bow lakes, far away in the swampy grass plains towards the Gulf of Papua to the south.

With such visions in my head one morning, I met Irari coming from the forest with a branch from a *kapone* tree. This he began to work with the stone axe-adze assembled by Dogore, using a log as a chopping block. He cut off a piece 159 cm long, about his own height, and fashioned it into a heavy digging stick, a kind of lever with a diameter of 7 cm and weighing approximately 2 kg. At the lower end it was sliced away on one side to form a narrow blade. Subsequently Irari demonstrated how easily it could be driven into the earth (Fig. 48) and how effectively it could break the clay subsoil into lumps, which he threw away with his bare hands. In fact this was the way my cook had dug a drainage ditch around my house the day I arrived.

1. The Making of Sweet Potato Mounds

One afternoon Irari made a man's short digging stick from the same

Fig. 48. Irari breaks up the sod with a heavy digging stick at Puya-kira'go.

kapone wood as the heavy stick described above. In the local language it was called *kimbu* (?*kiembu*), was only 109 cm long and weighed 550 g. The maximum width of its blade was 4 cm (Fig. 49, bottom). This kind of digging stick is preferred by the men for garden tillage.

Irari also made a narrow spatulate tool, called *tsenge*, for his wife Pokome (Fig. 49, top). It was made of the same wood as the other sticks, and he roughed it out with the stone axe-adze previously made by Dogore (Fig. 50, cf. Fig. 11). The *tsenge* was nearly 82 cm long and weighed 250 g. Its upper end was 3 cm thick and almost oval in section, but at a distance of 15 cm from its working end it was 4·3 cm broad and only 2 cm thick. The front side was flattened and the rear side rounded. After it had been used for only one morning,

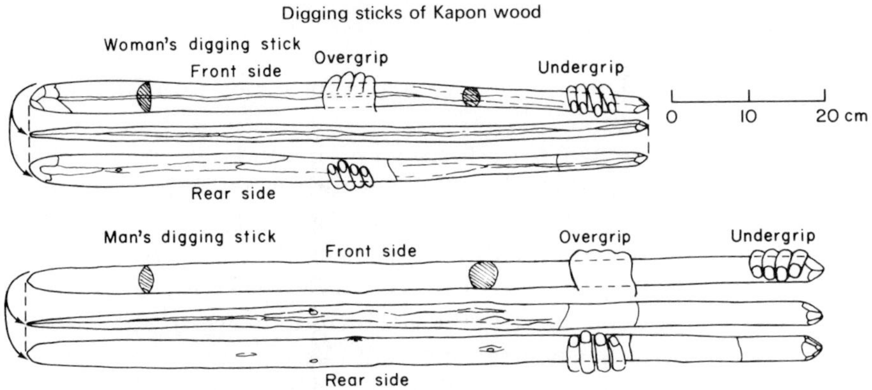

Fig. 49. A woman's digging stick (*tsenge*) above and a man's digging stick (*kimbu*) below made by Irari at Puya-kira'go. The tools are viewed from the front, from the side and from the back, the position of the user's hands being indicated. Drawn by the author.

Fig. 50. Irari roughing out the *tsenge* (Fig. 49, *top*) using the axe-adze made by Dogore.

it was frayed on its working edge and almost 0·5 cm of its length had been worn away.

One day Pokome had to re-form some mounds for sweet potato (*Ipomoea batatas*) in the deep valley south-west of Puya-kira'go. They were in a small garden belonging to her husband but tilled by another woman, and it had recently been destroyed by feral pigs breaking through the fence. In Fig. 51 Pokome is shown pulverizing the soil, grasping her stick with the left hand in overgrip about 39–48 cm from its working end and the right hand (not visible on this picture but indicated in Fig. 49, top) in undergrip around the upper end of the stick. She also used her hands when clearing the soil and producing a good tilth.

The distance between the hands varies according to the kind of tool and the nature of the work going on. When Irari demonstrated the use of the *kimbu* he pushed it into the grass soil as illustrated in Fig. 52, but afterwards when he pulverized the soil he had dug up, he used it in the same way as his wife used her *tsenge*. The heavy lever-stick shown in Fig. 48 was lifted with both hands placed rather close to each other, but when used for breaking clods of earth the right hand was pushed upwards grasping the end of the stick. With a shorter and lighter stick the hands must be placed closer to each other when breaking the soil, and when it is used for pulverizing, the lower hand must be placed close to the working end because of the resistance of the soil.

The two men from Alipe, observed in 1968 (Lerche and Steensberg, 1973) placed their hands at an even greater distance than Pokome of Puya-kira'go because their spades had broader blades. Consequently to use these spades required greater effort than the narrow spatulate stick Irari had made for his wife. As well as their digging tools these two men also used their hands as claws, especially when clearing the soil in exactly the same way as Irari's wife.

At Yuku, Kualu's wife, Dayame, made five or six nicely rounded and vaulted mounds as shown in Fig. 53. Further away behind a heavy tree trunk, older mounds can be seen densely grown with sweet potato vines, alongside two or three cultivated *Pandanus* trees. In the foreground and upper right are old sweet potato beds, almost disused and overgrown with weeds, and some sugarcane (*Saccharum officinarum*). Further uphill, some taro (*Colocasia esculenta*, in the local language called *la*) were planted in the fallow fields. According to my interpreter, sugarcane—called *angu*—is cultivated by women, whereas bananas (*Musa* spp.), called *hale*, are mostly cultivated by

Fig. 51. Pokome, Irari's wife, pulverizes the soil with her *tsenge* to make a sweet potato mound below Puya-kira'go.

Fig. 52. Irari demonstrating the use of the *kimbu* (Fig. 49, *bottom*).

Fig. 53. Kualu's wife, Dayame, has completed some sweet potato mounds at Yuku near Puya-kira'go and planted them with fresh vines. The mounds behind the trunk are covered with developed vines.

men. Yams (*Dioscorea* spp.), called *ere*, are also mostly cultivated by men, but at Puya-kira'go and Yuku the altitude is too high for producing good crops of this plant.

In New Guinea the cultivation of sweet potatoes is normally women's work. However, in Alipe in the Kaugel valley half the male population are unmarried, because they cannot afford to buy a wife, and these, together with some elderly widowers, have become clever cultivators of this crop. Grith Lerche and I observed two men, Kupala and Wapea, making a new sweet potato mound by breaking up an old mound which had been fallow for about eight months. In other cases the fallow period was two to three years.

The two men worked as a pair, and in this respect it was an advantage that Wapea was born left-handed. They started from the

Fig. 54. At Alipe in the Kaugel valley Kupala breaks up the soil of an old sweet potato mound with short, rapid strokes. His left hand grasps the short paddle spade in overgrip at the top of the blade thus exerting the maximum force.

top of the fallow mound and worked downwards, breaking the black volcanic soil towards themselves. In Fig. 54 Kupala is shown putting his spade in from the right and throwing the soil to his left with small, rapid strokes. He broke up the lumps with the edge of his spade and pulverized the soil with his fingers, throwing waste material into the centre of the mound.

Wapea worked in the same way but in the opposite direction, his right hand holding the spade very close to the blade so that he could exert his maximum strength (Fig. 55). Kupala used his left hand for pulling the spade towards him and controlled it with his right. Both men worked strenuously in a squatting position, and sometimes they scooped the loose soil uphill (Fig. 56). Children were frequently interested on-lookers, learning the technique indirectly. Grass roots found in the soil were sorted out and thrown into the centre of the

Fig. 55. Kupala (*left*) and Wapea (*right*) of Alipe working together at breaking up an old sweet potato mound. Wapea was left-handed.

Fig. 56. Working in a squatting position Wapea scoops loose soil into the centre of the mound with his hands.

mound. Old sweet potato vines were separated out and piled outside the mound; a few of them proved to be fresh enough to be replanted, but most would be burned later on. The presence of charred vines and stalks between the mounds provided evidence of this practice. Most vines for replanting were taken from a mature, flourishing mound in another part of the garden, probably a special variety which they liked to grow again.

When the sides of the mound had been carefully worked all the way round, the two men brought large bundles of old dry vines and placed them in the middle of the mound. Then they scooped the earth up with their hands, taking only 2 or 3 minutes to cover this compost layer. Seven to eight fresh vines were laid out on the surface of the finished mound, and the lower ends of these were pushed into the loose soil so that only the tops were visible. Some more soil was then scooped up using their hands, and if necessary the lower fringes of the mound were cleared with the spades, loosening more soil which could then be scooped up. If the tops of the vines happened to be covered by earth, they were pulled free, and occasionally the men patted the soil between the vines with their hands, but not so much that the soil was compacted. The surface was left crumbled and porous.

Manure is never used directly. Usually the faeces of the pigs are thrown out of the stalls and gradually fertilize the area around the houses (Wirz, 1924, p. 98). However, in Puya-kira'go, people told me that they generally left the faeces in the stalls, so that when they moved house three to four years later the material remaining rapidly decomposed and was taken up by the soil. People made their toilet in the bush, and over a period of time the soil around the settlement was fertilized. According to Clarke, the Bomagai-Angoiang had special areas near their settlements, one for each sex, where the inhabitants discharged faecal matter and urine. When one area had been used for a while, it was abandoned in favour of a fresh place. Later, when the faeces had decomposed, the area was considered to be a good garden site and cleared for crops (Clarke, 1971, p. 72). Conscious and intensive manuring like that practised in ancient China finds no place in the New Guinea Highlands, partly perhaps because of lesser pressures on the land, partly because of different techniques of working it.

Mound cultivation of the sweet potato is characteristic of high altitude areas to the west of Mount Hagen. It has been described by Freund (1968) who lived at Mukuramanda in the Lai valley between Wabag and Wapenamanda, and is also the subject of a monograph by Waddell (1972) who worked in the same area. Freund describes how the Enga people make mounds with diameters ranging from about 1·5 m to 4·5 m and heights from about 0·45 m to 1·20 m, depending on the area and soil type. He summarizes the advantages of this type of cultivation as follows:

Drainage: Tubers and plants in the mounds are never waterlogged. Compaction of soil is minimized.

Depth of soil: Where the soil is shallow, mounding increases the effective depth of soil.

Fertility: Composting returns some nutrients to the soil.

Frost protection: Composting raises the temperature within the mound which itself raises the level of most of the vines above the coldest air layers, thus minimizing frost damage.

Plant population: Is reduced by mounding, thereby enabling availability of plant nutrients and production to reach equilibrium.

Photosynthesis: Mounding increases the effective surface area of soil over which vines can take place [sic] on a given area.

Mounds are usually laid out on sloping ground in such a way that a lower mound faces the opening between two mounds above it. On steep slopes the mounds tend to act as a kind of terrace, protecting the soil from being flushed down by heavy rain. Branches and trunks of trees may also be placed amongst the mounds to check the flow of water. Waddell quantifies some of the observations made above. He particularly stresses the importance of mounding as a defence against frost, and in a more recent paper (1975) describes how this forms only one of a complex of strategies employed by Enga tribesmen living above 2250 m altitude to deal with an ever present natural hazard.

2. Flat-Bed Gardens

Powell *et al.* (1975, p. 9) describe another form of intensive husbandry characteristic of the Mount Hagen area. The sweet potato gardens of this district consist of flat square plots (*mai oka* in the local language), mostly cleared from grassland on not very fertile soil. The square plots are divided from each other by 30 cm wide ditches from which the soil is thrown up on to the beds, which are about 30–45 cm higher than the bottom of the ditch (Fig. 57). After it has been tilled for four to six years with only four to six months intervening fallows, the "hard" ground will become "soft". After this it may either be reworked into mounds (*oka mund*) or left to fallow for 20–30 years. These *oka mund* are themselves recultivated in one of two fashions, one being simply their reshaping in the manner described for Alipe. In the other method the weeds and grasses are pulled off mounds after fallowing and heaped up in between. An old mound is then dug into and the soil from it thrown on top of the heaps of compost on each side. Thus each time a garden is reworked, the mounds are displaced (Fig. 58), so that where there were spaces before, there are now mounds (Powell *et al.*, 1975, p. 11).

The tool for trenching in dry-land gardens seems to have been a double-bladed paddle spade. The two old men from Mbukl, El and Ul, made some replicas of this type, and in a garden about 2 km west of Kuk Tea Research Station El demonstrated how to form the flat square beds for sweet potatoes with such a spade. He measured the plot out with a rod, and along this he cut a straight line (Fig. 59). Afterwards he dug the trench (Fig. 60) and I saw that he used one end of the spade for digging and the other for shovelling. He also removed loose earth with his hands, scooping it up on to the bed. When I tried to dig I noticed that the end he used for shovelling was about 1 cm broader than the other, though he claimed that they were both equally long and broad.

Old men of the district told me that their fathers used to say they felt that their effort was eased by turning the spade and using the other end. This is possibly true. Ongka of Mbukl, chief informant of Andrew Strathern, an ethnographer of the Mount Hagen people, suggested that there might be a psychological element in the use of the double spade, which was a tool he remembered well. However, another explanation may be more relevant. A wooden spade soon becomes worn, as I noticed at Kuk, and then it needs to be re-sharpened. When the men went out into the fields, it was usual for them to take with them a bow and a dozen or so arrows as well as their spades, and therefore to carry a heavy stone axe or adze in addition would be cumbersome. The use of the double spade reduces this necessity at least by half.

This explanation is supported by examples from other parts of the world, where such double spades were also formerly manufactured and used. Lerche recorded 280 finds of double spades from peat bogs, nearly all in Jutland (Lerche, 1977), and two of them

Fig. 57. Square plots for sweet potatoes (*mai oka*) in the Wahgi valley, 1968. The soil from the grid of trenches is thrown up to form the garden surface, which will be for planting.

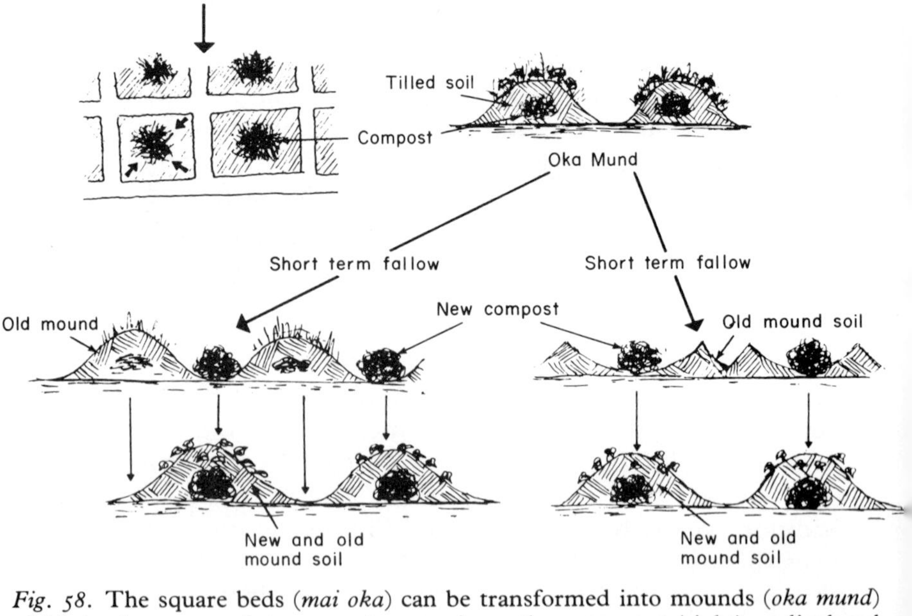

Fig. 58. The square beds (*mai oka*) can be transformed into mounds (*oka mund*) which in the course of time may be remade on the same spot (right) or displaced to fill a previous gap (left). The former is the way it was done at Alipe (Figs 54–56). After Powell *et al.* (1975), with permission of the Department of Geography, University of Papua New Guinea, Port Moresby.

Fig. 59 (*left*) El from Mbukl making a *mai oka* bed in Mak's garden at Kuk using a double paddle spade made by him after the traditional model. He cuts a line along a stick which marks the border of the bed. *Fig. 60* (*right*) El digs the garden trench with his double spade.

were dated to about 2200–2300 years ago. Some of them had been worn at both ends and there were traces of resharpening. One of them is drawn in Fig. 61. They were possibly used for digging trenches like the New Guinea spades.

According to Raum (1977, p. 100), the Xhosa of south-east Africa used a spade made from one piece of wood with a blade at either end of a common handle. Also, some of the Zulu farther north used such a tool, and Raum quotes Alberti (1968) for a description of the fire-hardened "spadoon", made from one piece of wood, in such a way that at both ends of a common handle, which was 2 in. (5 cm) thick and rounded, "a blade of about four inches [1 cm] in width and ten inches [25 cm] in length is formed" (Raum, 1977, p. 99).

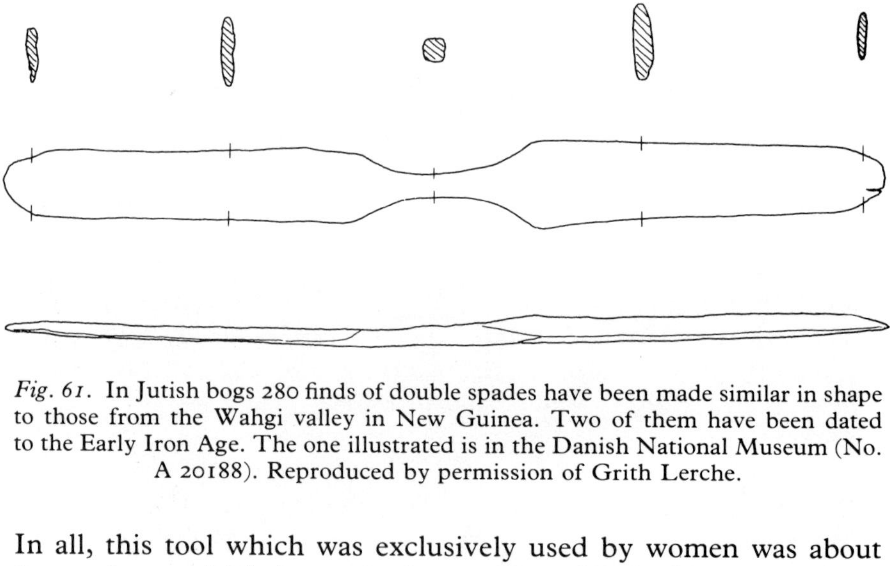

Fig. 61. In Jutish bogs 280 finds of double spades have been made similar in shape to those from the Wahgi valley in New Guinea. Two of them have been dated to the Early Iron Age. The one illustrated is in the Danish National Museum (No. A 20188). Reproduced by permission of Grith Lerche.

In all, this tool which was exclusively used by women was about 60 cm long, which is much shorter than the double spades from Jutland and New Guinea. There is, incidentally, another parallel to the practice of tilling in the New Guinea Highlands amongst the Zulu of South Africa who, like the Enga, used mulching as a substitute for manure. The examples mentioned are striking proofs of how people in different parts of the world have developed similar practices independently as a response to the environment. They have simply adapted their basic technical inventiveness and skill to similar needs.

The intensive practices we have described for the Enga and Mount Hagen areas are applicable to the cultivation of the sweet potato. But in both these areas there are also mixed gardens containing sweet potato, sugarcane, edible pitpit (*Saccharum edule*, but especially *Setaria Palmifolia*), taro, yam and banana. Such mixed planting is described in detail by Clarke (1971) for the Bomagai-Angoiang, and by Aufenanger and Höltker (1940) for neighbouring parts of the Bismarck Range.

Pandanus may be planted after the first crop of tubers is harvested or they may form part of an orchard around the house, together with banana, sugarcane, breadfruit, bamboo, cordyline and casuarina. Tobacco, beans, gourds, pumpkins, cucumber and other green vegetables, some introduced by Europeans, are also cultivated round houses. Above 2100 m altitude, *Setaria palmifolia* has probably been important from the earliest times and was presumably first domesticated in New Guinea.

3. Water-Controlled Gardens on Riverine Plains

In the flat-bottomed Grand Valley of the Baliem river, the Dugum Dani have cut a labyrinth of ditches which serve both to drain and irrigate the garden plots which flank them. Logs span the ditches and serve as footbridges. Perhaps half of the land lies fallow at any one time, and during the fallow period the banks of the ditches break down and the debris settles in the mud. Before the cultivation of the plots can be repeated, the men and boys have to go down into the deep ditches and using fire-hardened paddle spades cut out lumps of fertile mud, which they throw on to the garden beds on both sides, to be spread evenly over the garden surface by other men and women with their feet. Frequently a man will set his spade aside and reach into the ditch with his hands to grasp smaller lumps of mud, which he plasters against the sides of the reshaped canal, similar to the practice of the Indonesians in fashioning the terrace walls of their paddy fields. According to Gardner and Heider, the mud soon bakes dry in the sun and gives these gardens a look of permanence, which lasts until they are abandoned to fallow again (Gardner and Heider, 1968, p. 40).

The natural fertilizing effect of mud has been known from the very earliest times, for example, the Nile valley, and mud has been extracted from canals and village ponds in most parts of the world. Many years ago the late palaeobotanist, Johannes Iversen, and I excavated the pond belonging to a farm in Jutland (Alstrup Heath), abandoned in late medieval times, and it proved to have been emptied of muck and debris at least once, if not repeatedly. In 1822 the owner of a Jutland manor, Herningsholm, used the mud from the collapsed moats of the fortification to halt the drifting sands of his cultivated fields, which were suffering from a severe summer drought.

The Dani women plant sweet potatoes in mounds about 40 cm high, scraped together with flat digging sticks. They also plant the tops of taro tubers in shallow holes in the ground. Primarily this is done on the edge of the ditch, above water level; but some taro may be planted in the mud of the ditches themselves, because there are varieties that grow in wetter conditions. Mud is also spread over recently planted mounds, and when it has dried and cracked, the soil is broken up with a light digging spade (Heider, 1970, p. 40).

Le Roux had observed the same kind of ditches in the 1930s, when he visited the Baliem valley. They were up to 2 m wide and 2 m deep, and the mud from their bottoms was thrown on to the cultivated beds as described by Heider. Le Roux also relates that the Ekari at the Wissel (now the Paniai) Lakes, made drains with wooden spades and these led into deeper canals around the gardens, 1 m deep or more, from which the water was led to small streams and morasses near the surface of the lake. He described the digging stick as a spatulate tool, about 75 cm long, with a narrow blade with two sharp edges ending in a point, quite like the spatulate stick from Puya-kira'go, described above. The tool was pushed into the soil and a lump carved out in a circular movement, which was then removed by hand (Le Roux, 1948, pp. 289–291).

From Heider's pictures (1970, p. 41) as well as from photographs in Gardner and Heider's book (1968), it appears that the Dani had heavy lever-sticks, which were used by teams of men to break up grassy ground. Long broad-bladed digging sticks were adapted for work in deep muddy ditches. Similar spades have been found in great numbers in the swamps around the upper Wahgi river. One of them was excavated at the Kuk swamp in my presence, and afterwards dated in the Carbon-14 Dating Laboratory of Copenhagen to about 500 years ago. Some years ago a woman's short digging stick was found in a nearby swamp, dated as early as 2300 years ago (Lampert, 1967).

The archaeological investigations in the upper Wahgi valley are providing a wide perspective in the development of water-controlled agriculture in New Guinea (Golson, 1977b). This possibly began as early as 9000 years ago, in association with the first clearance of forest for agriculture in the catchments of the Kuk swamp. Starting

6000 years ago there is clear evidence for different episodes of swamp management, ending about 100 years ago. The drainage systems which are operating today at the Paniai Lakes in the Baliem valley in Irian Jaya are adapted to the cultivation of the sweet potato, a water-intolerant plant. The regime of swamp agriculture at Kuk seems to have been remodelled about 250 years ago to make provision for the sweet potato, which is thought to have entered New Guinea only shortly before this date. Before this time crops like taro, yam and banana must have been staple foods in the Highlands.

Golson describes the digging of large drains across the swamp about 6000 years ago, some 2 m wide and 2 m deep, all of them well over 1 km long. Field systems associated with these drains consisted of "a web of short channels, so disposed as to define roughly circular clay islands of about a metre diameter" (Golson, 1977b, p. 616). Golson supposes that these islands, elevated by the soil from the channels, were planted with less water-tolerant crops like yam and banana, while the water-tolerant taro was grown in low basins at the intersection of the channels. The system operated as required for drainage and for irrigation. Subsequent systems, at about 4000 to 2500 years ago, 2000 to 1200 years ago and 400 to 250 years ago, are characterized by increasingly elaborate drainage works, as man's interference with the natural drainage regime made the area progressively more swampy. In these circumstances Golson supposes that cultivation concentrated more and more on taro.

The construction and maintenance of ditching systems on this scale must have required longer spades than those used for digging the small trenches between the square dry-land sweet potato plots, described above. The longer spades, up to 3 m in length, are commonly found when the Wahgi swamps are drained for agricultural development today.

4. Demonstrating the Use of Wooden Digging Tools at Kuk

Golson and I thought we might learn something of the qualities and effectiveness of such wooden tools if we could use the living tradition embodied in the old men from Mbukl, El and Ul. Consequently we devised a simple demonstration which, although it could not be described as scientifically controlled, proved instructive in assessing the performance of wooden spades and comparing them with steel spades. In considering the results, there are some important points to bear in mind.

(1) Wooden spades have not been in use in the Mount Hagen area for many years. Though older men remember them well, the younger men employed by Golson on his archaeological project who were to perform the demonstration, were accustomed only to steel spades. There was a limited amount of time to have them instructed by the older men in the use of the wooden spades. At first they were all keen to try, but the enthusiasm of some waned in the course of working for long periods with such unfamiliar tools. Besides El, who despite his greater age insisted on taking part in the demonstration, Ru, a man of authority with an intellectual interest in traditional ways, was important in keeping the spirit of his fellow workers from flagging too much.

(2) There was a problem with the wooden tools themselves. Old spades found in the upper Wahgi swamps during the course of drainage are almost invariably long-handled, single-bladed specimens, sometimes with the upper part of the handle left thickened to counterbalance the blade. Few of these genuinely old tools were for obvious reasons available or fit for use in the demonstration, but we did have access to specimens found some years previously during initial drainage of the Kuk swamp, put away in the Tea Research Station store and still sufficiently well preserved to be used. In order to supplement these, we asked Ul and El to make replicas of wooden spades for use in the demonstration. However, they supplied not only single-bladed specimens but also double-bladed spades of the type described in section 2 above in connection with the making of dry-land sweet potato beds.

(3) A third question involved the best way of comparing the performance of wooden and steel spades, especially in the light of the considerations discussed above. In organizing labour for drain digging during his excavations in the Kuk swamp, Golson adopted the practice common on local plantations of setting each workman a quota for a day's work, on completion of which he was free to go home. The workmen themselves liked this practice since by working hard they could finish early and so have time to do necessary work of their own in their gardens and households. It seemed to us best to retain this incentive in the proposed demonstration. However, in order not to give an unfair advantage to the men working with the steel spades, the daily quota was halved for the men using wooden spades. Thus the latter each took a 5 m length of drainage

ditch, to be dug to a width of 1 m and a depth of 1·6 m, while the men with steel spades took the usual length of 10 m. These dimensions were set by the fact that the ditch had to fit in with the Tea Research Station drainage plan and so had to be dug to its specifications. It should be noted, however, that in the digging of such drains the walls were not dug vertically but battered inwards towards the base. There could be much variation between individual workers in the degree of batter they make and therefore in the amount of earth they removed.

(4) Following the advice of El, we provided each user of a wooden spade with a helper to assist in raising the earth out of the ditch by steadying the end of the wooden blade with its spoil and, with the digger holding the handle, lifting it up and unloading it over the side of the drain. This would seem to be an efficient method where the earth retains its cohesiveness. Unfortunately the ground at the drain we chose for our demonstration had been affected by drying out due to the beginning of the drainage of the Kuk swamp some years previously, so that the top 50 cm or so was harder than desirable and crumbled under the impact of the spades. This doubtless affected the performance of the men with steel tools, since they were using spades, not shovels with a retaining flange. It affected the users of wooden spades even more, however, because of their narrower blades. As a result the assistants of these men had no real utility for long periods, during which their lack of employment doubtless interfered with the concentration of their co-workers. Only in the bottom of the drains, where the clay was still wet, could they perform their appointed role with anything like efficiency.

To prepare the demonstration, the men cleared the growth of 4 m tall cane grass and the underlying mat of shorter grasses, partly using steel bush knives and partly *yakla* knives of palmwood, as described in a previous section (Fig. 37). This was followed by a day or so when the men were instructed by El in the use of wooden spades, which was sufficient at least to prove the adequacy of both the swamp-found and the manufactured examples. In the foreground of Fig. 62 Ru is seen holding one of the replicated spades and discussing the planning of the demonstration with Golson. This spade was made of casuarina wood and quite heavy. It was about 185 cm long, with the blade about 12 cm wide and 50 cm long.

Then the men set two strings, nowadays of store-bought twine, but traditionally made from forest creepers, in order to cut the ditch in a straight line (Fig. 63). In the centre of the picture El and Ul (behind the stake) lift their long paddle spades (both of them swamp finds) and jab them into the tough surface. To the left of them Ru cuts a straight line for the edge of the ditch. Both El and Ul use the forehand (left) in overgrip and the rear hand (right) in undergrip, whereas Ru uses his right hand in overgrip and his left hand in undergrip because of the different nature of the work. Ru's spade is seen edge on; it is shorter than the other two but is thickened at the upper end in order to counterbalance the long blade. Figure 64 shows the work in progress with several men digging the demonstration trench with 3–4 m tall pitpit in the background.

In Fig. 65 El is shown digging near the bottom of the demonstration trench with one of the spades he had made, the handle of which he had left rather heavy at the top in order to counterbalance the blade; many of the fossil spades found in the swamps were made in this way, sometimes similar to a second blade but with blunt edges. Most of the time El used a double-bladed paddle spade as shown in Fig. 66, his left hand in overgrip. When he had dug deeper in the ditch, he often grasped the end of the upper blade with his right hand, pushing the tool into the wet clay. He was able to cut out slices of clay about 30 cm long and 8–10 cm wide, lift them from the bottom of the ditch and throw them on to the bank, where they soon dried in the heat of the sun (Fig. 67).

A full description of this undertaking will be given in Golson's report on the Kuk investigations. For present purposes it is enough to note that on the first day of the two-day demonstration four teams of two men, a digger and his assistant, completed 5 m of drain each in an average of 3 hours 41 minutes (range 3 h–4 h 15 min), while nine individuals with steel spades each dug 10 m of drain in an average of 5 hours 20 minutes (range 4 h 20 min–5 h 50 min). The enthusiasm of most of the two-man teams using wooden spades waned on the second day. However, by this time it was evident that spades used for digging deep drainage ditches ought to have a long shaft. El said he found the double-bladed spade less convenient when used in the bottom of the ditch and admitted that its advantages lay especially in digging the shallow trenches around the sweet potato beds. Indeed, in the bottom of the ditch he used a single-bladed spade, as shown in Fig. 65. It is not therefore surprising that the vast majority of the spades coming to light as a result of drainage of the Wahgi swamps are of the single-bladed type with a long handle.

A double spade from the Kuk swamp has recently been published by Gorecki (1978); it was found in a native garden close to the swamp margin at a depth of 75 cm in the fill of an old ditch. Gorecki suggests it was used for digging in the swamp, but according to the arguments advanced here it would have been used for tilling sweet potato beds on the nearby dry land and temporarily lodged in the ditch to keep the timber moist.

I was able to make a Super 8 mm film of our demonstrations with

Fig. 62. In the swamp at Kuk Tea Research Station the ground has been cleared for the digging of a demonstration trench with wooden paddle spades. Jack Golson discusses the organization of the work with Ru, who is holding one of the spades to be used.

Fig. 63 (*left*) The demonstration trench at Kuk has been marked out with sticks and ropes and the men are beginning to define its edges and open it up. *Fig. 64* (*right*) Digging the demonstration trench at Kuk. The men in the foreground are using wooden spades, the others beyond are using steel spades.

Fig. 65 (*left*) El digging about 1·5 m deep in the Kuk demonstration trench using a long wooden spade. *Fig.* 66 (*right*) El digging in the demonstration trench at Kuk using a double-bladed wooden spade which he holds with his left hand in overgrip.

Fig. 67. The slices of clay produced by the wooden spades were rectangular, about 30 cm long and 8–10 cm wide.

Fig. 68. The lumps of clay produced by the steel spade were roughly square, but Wai handles the spade in the same way as a wooden spade with his forehand in overgrip.

the result that future research can profit from the experience of an old man knowledgeable in the use of wooden spades. It is interesting to note that some of the men using steel spades handled them in the same way that El handled his wooden spade, with their left hand in overgrip (cf. Fig. 68 with Fig. 66). The steel spade removed a wider and thicker slice than the wooden spade (cf. Fig. 68 with Fig. 67).

5. A Survey of Gardening and Trenching Tools

The aim here is simply to bring together the different sorts of agricultural working tools which we have described in use in different parts of the New Guinea Highlands as a basis for some comparative remarks. Though the forms reviewed are called types, it is as a matter of convenience: there is no intention of setting up a typology to integrate the various aspects of technology, functions, association with the sexes and geographical distributions. A similar exercise has recently been attempted by Golson (1977a).

Type 1: digging sticks bevelled by one or more facets on one side at one end to produce a rounded point. A distinction is made between shorter (a) and longer (b) examples.

(a) This is the universal woman's tool, used for a variety of purposes in gardening, including harvesting, weeding and loosening the soil. Figure 69 (1)–(3) are examples in use when collected in the Mount Hagen area. Figure 69 (4), another ethnographic specimen, is a *kope* from Alipe, actually a ceremonial digging stick, with one end bevelled and the other pointed. Figure 49 (top) a *tsenge* from Puya-kira'go, made by Irari for me (Fig. 50), is also of Type 1a and is used by the women of the district in reworking the sweet potato mounds (Fig. 51).

(b) This tool is normally a metre or more in length. Figure 69 (5) is a *tsu* from Alipe, an ethnographic specimen, and is used for re-forming sweet potato mounds. Figure 49 (bottom) a *kimbu* from Puya-kira'go, made by Irari for me, is used by the men (cf. Fig. 52), amongst other things for their work in the mixed gardens. Figure 70 (8) is a fossil example of a similar type found in the Kuk swamp.

Type 2: a heavy digging stick used by the men for breaking ground, as a lever for turning over soil, and similar tasks. Figure 48 shows one in use by Irari at Puya-kira'go but it is a widely distributed implement.

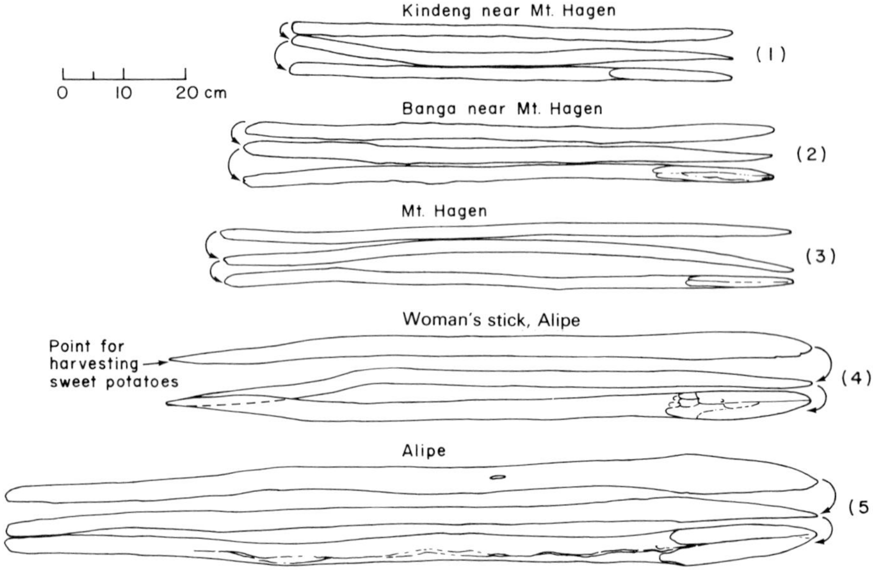

Fig. 69. Five digging sticks. (1)–(4) are of type 1a: (1) from Kindeng, (2) from Banga, (3) an *øpukl*, from Mount Hagen—all in the Hagen area and (4) a *kope*, from Alipe (a woman's ceremonial stick). (5) of type 1b is a *tsu* from Alipe.

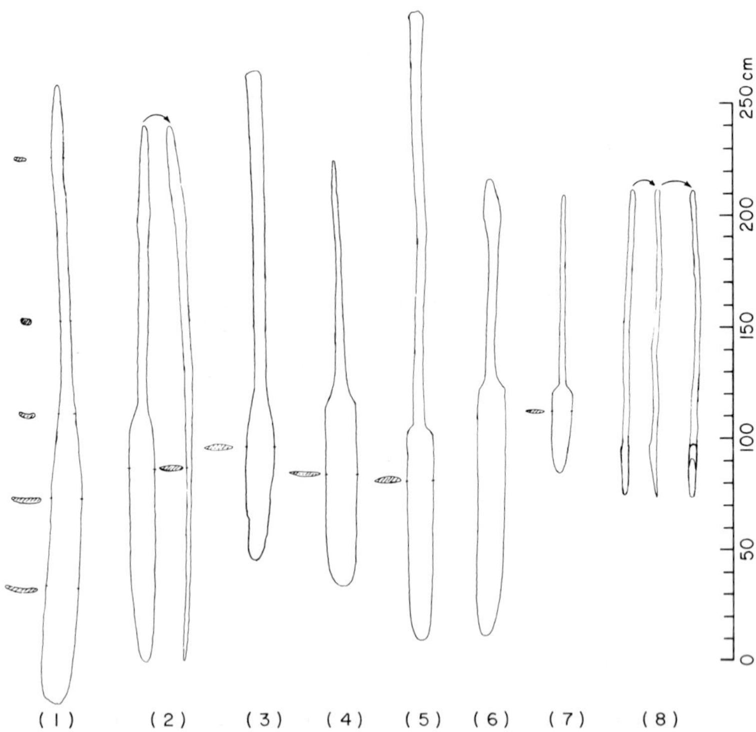

Fig. 70. Seven paddle spades and a digging stick found in the swamps of the Wahgi valley near Mount Hagen: (1)–(6) are of type 3b, (7) of type 3a, (8) of type 1b. Figs 69 and 70 drawn by the author.

Type 3: paddle-shaped spades, divided into shorter (a) and longer (b) examples.

(a) Figures 54–56 show two examples in use in reworking sweet potato mounds at Alipe. Figure 70 (7) is a small fossil spade from the Kuk swamp, of unknown use.

(b) A long-handled paddle spade adapted especially to the digging of deep drainage ditches in swamps and used by men. Figure 70 (1)–(6) are examples recovered from swamps in the upper Wahgi valley but the type is known from other areas of the New Guinea Highlands.

Type 4: a double-bladed paddle spade used for digging shallow trenches round square garden beds, drainage ditches round houses and along fences. It is known from the Mount Hagen area, and Figs 59–60 show El using an example he had made for us. Figure 61 shows similar implements from prehistoric Jutland.

Type 5: a short-handled paddle spade with sharp shoulders, called a "hastate spade" by Powell (1974) who says it was used in the Mount Hagen area for marking out plot lines, cutting surface turf and making small holes.

In 1968 Grith Lerche and I recorded a collection of seven paddle spades (type 3b) and one digging stick (type 1b) found during drainage of swampland at Kindeng down the Wahgi valley from Kuk and housed in the Department of Forestry office in Mount Hagen (Lerche and Steensberg, 1973). In 1975 only two of these spades were extant, being then part of a small museum display at the Mount Hagen High School (Fig. 70 (3) and (6)); perhaps the others had disintegrated through lack of conservation treatment. These objects were remeasured in 1975, together with four paddle spades (Fig. 70 (2), (4), (5), (7)) and one digging stick (Fig. 70 (8)), stored at Kuk Tea Research Station since the time of their discovery during drainage of the Kuk basin. The measurements of all the above, together with those of the 500-year-old spade (Fig. 70 (1)) found during the digging of the demonstration trench described in section 4, are given in Table II.

The blade-shaped upper end of (6) is not an exceptional case. In Fig. 71 the upper ends are shown of two other spades, both of which have a distinct triangular blade opposite the paddle-fashioned digging blade. These upper blades are about 25 cm long and 8 cm

Table II

No. in Fig. 70	Full length (cm)	Approximate length of blade (cm)	Breadth of blade (cm)	Additional remarks
(1)	284	130	15	Found at Kuk 3 October 1975. Dated in the Copenhagen Radiocarbon Laboratory to about 500 years ago (Laboratory no. K 2643).
(2)	242	115	11	Found at Kuk. The upper end is fashioned as a counterbalance.
(3)	220	75	12	Found at Kindeng, previously in the Forestry Office at Mount Hagen. The upper end is slightly thickened.
(4)	192	88	13	Found at Kuk. The upper end tapering and unusually thin.
(5)	283	96	12	Found at Kuk. The upper end thickened.
(6)	206	115	12	Found at Kindeng, previously in the Forestry Office at Mount Hagen. The counterbalance is in the shape of a blade at the top end, 8 cm broad with blunt edges, 23 cm long.
(7)	125	40	10	Found at Kuk. Very small: Type 3a.
(8)	138	15		Digging stick found at Kuk. Type 1b.

All implements above are of type 3b except where otherwise stated.

wide, but the edges to the right are blunt (about 1 cm thick), and so are the straight edges to the left (about 1·5 cm thick). The paddle-shaped working blade at the other end of the specimen on the left in Fig. 71, which was turned about 35° in relation to the plane of the upper blade, seemed to be slightly resharpened on its right edge when seen from the flat side. This would presumably be the leading edge when cutting a straight line.

Of course there would be an ideal point of balance between the blade and the handle in order to make such tools function well in the hands of the user, just in the same way that its length should be adapted to his height. But as shown in Fig. 70 some blades were shorter than others as a consequence of resharpening, and some of

Fig. 71. The blade-shaped top ends of two paddle spades found in the swamps of the upper Wahgi.

the handles may also have been broken and shortened until the tool became useless and thus abandoned. However, most spades found in the swamp were most likely stored there when not in use in order to absorb moisture and recover their strength. Watson (1967, pp. 95–96) mentions this practice in eastern New Guinea, where people often used to leave spades in the bottom of a garden trench, from where they would be retrieved when needed. According to his illustration (Fig. 72), the spade of this region was furnished with a kind of hilt between the expanded oval blade and the straight round handle, while the end of the handle was formed like a sword-pommel. The lengths of the spades he measured varied between 103 cm and

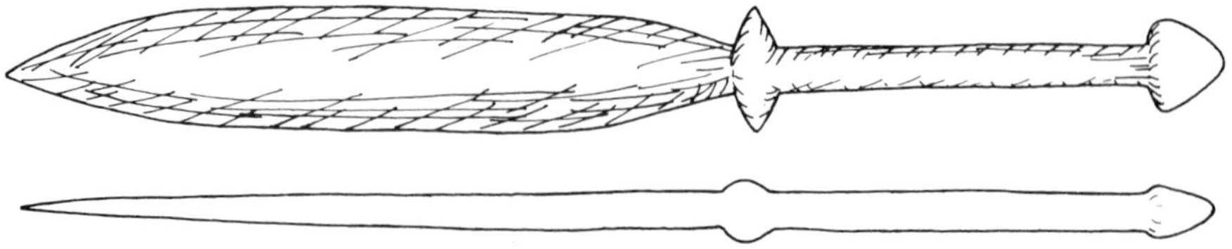

Fig. 72. From the Eastern Highlands Province of Papua New Guinea, Watson (1967, p. 96, Fig. 1) illustrates this sword-like wooden spade, shorter than the Mount Hagen spades. By permission of the editor of *Oceania.*

115 cm, the blades being 57 cm to 76 cm long, which is less than the majority of the Mount Hagen specimens. A casual demonstration of Watson's spatulate spade indicated that it was stabbed vertically into the ground with both hands and then levered backwards and forwards to loosen the soil or, to tear free a sod which then had to be lifted by hand. In fact these spades could not be used for digging deep trenches, but they were suitable for cultivation and digging garden trenches and surface drains.

However, they were larger than the short-handled hastate spades with the sharp shoulders illustrated by Powell (1974). These short-handled tools are similar to some short-handled shovels found by Schwabedissen in Satrup Moor, South Schleswig (Fig. 73a–c) and dated to about 6000 years ago (Schwabedissen, 1957–1958). Could the use of the hastate spades in New Guinea parallel the use of similar tools in the Early Neolithic of Northern Europe? Powell mentions that this kind of spade was used for marking out plot lines, for cutting surface turf, and for shaping and lifting material out from small holes (Powell, 1974). But short-handled spades were probably also well adapted to other kinds of small-scale scooping, and even for patting and beating the loose soil and smoothing the wall-surface of ditches.

In prehistoric Europe there is a distinct structural difference between such straight board-like tools with the handle cut out in the same plane as the blade and with normally a rather rounded cutting edge (Fig. 74a), and some other short-handled tools in which (as in the long-handled shovel from Satrup Moor) the front side of the triangular blade is almost a continuation of the back side of the shaft, i.e. the shaft continues as a short ridge on the front side of the blade (Fig. 74b). Two blades of this second type were found south of Århus, but details of them have not yet been published. They are dated to the same Early Neolithic period as the Satrup

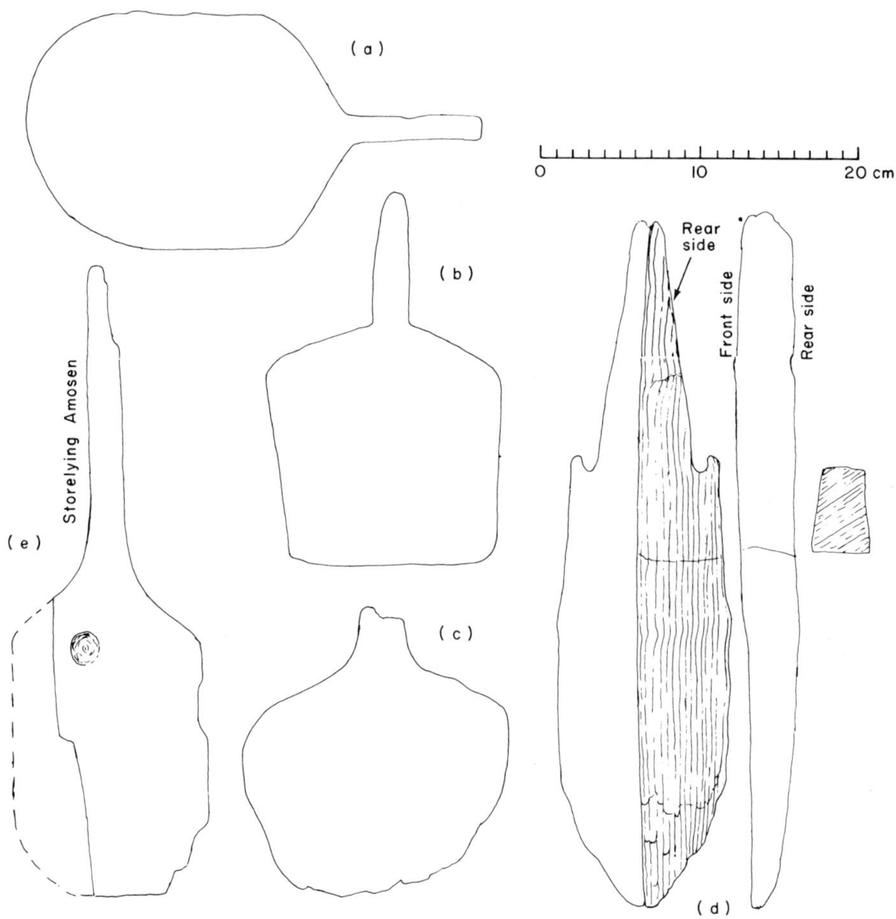

Fig. 73. Four Early Neolithic wooden tools from Satrup Moor in South Schleswig (a–d) and one from Åmosen in Zealand (e). Drawn by the author by permission of H. Schwabedissen and J. Troels-Smith.

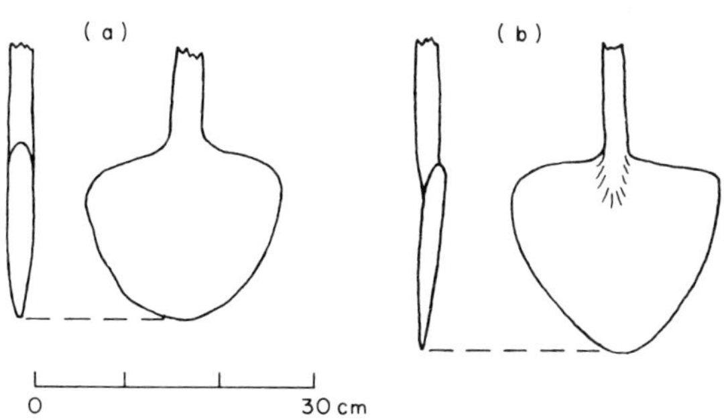

Fig. 74. Two types of "spade" from the Early Neolithic of Jutland, both made from a single piece of wood. On the left the blade is a direct continuation of the handle, on the right it is offset. Drawn by the author.

spades, and they are now kept in the Forhistorisk Museum at Moesgård, Århus.

Another Early Neolithic tool was found in Åmosen, Zealand (Fig. 73e). This was originally supposed to be a paddle, but an attempt had been made to drill a hole in a similar position relative to the shaft as one of those made on a tool from Satrup Moor interpreted by myself as a rope traction ard (Steensberg, 1973, p. 105). However, the hole had not penetrated the blade, and it is assumed that a similar hole symmetrically to the right of it was planned but had not been begun. Also the shaft was a straight continuation of the blade, like Fig. 74a, not offset as the blade in Fig. 74b. Because the lower edge seems to be cut square in relation to the sides, it is more likely to be a short hastate tool similar to Fig. 73a, b and c, and to the short hastate spades from New Guinea.

The tool shown in Fig. 73d was found on Satrup Moor in the same layer as the three short-handled spades. It seems to be one half of a symmetrical tool which had a long shaft and was furnished with rounded notches on each shoulder in which a traction rope could have rested. Consequently this tool demonstrates another type of rope traction ard, differing from the implement with two perforations in the top of the blade published by myself (Steensberg, 1973), though both types were pulled through the soil like an ard-plough. The tool shown in Fig. 73d was stronger and narrower, more suited to deep cultivation of the soil than the tool with the two perforations through its triangular blade. The rear side of its point was worn to an angle of about 30° with the front, as can be seen from the side view, to the right. It must be compared with the ards for breaking the ground known from Døstrup in Jutland (Glob, 1951, p. 36) and from Israel (Avitzur, 1965, Fig. 22).

The examples mentioned illustrate the great variety of tools available for the cultivation of the soil in Neolithic times. Müller-Beck (1965, pp. 58–62) illustrates a similar range developed in the Swiss Lake Dwellings. Though he talks cautiously of *gestielte Blätter*, blades with a handle or shaft, he is inclined to explain their functions as *Grabstöche*, digging sticks, or *Grabscheite*, spades. One example seems to be too weak to have been used in the soil, while some were probably shovels because their long narrow blades had a square cutting edge, quite similar to two specimens found at St Aubin and Port in France, thought to belong to the Cortaillod culture (Müller-Beck, 1965, Figs 120–122). Another specimen was found at Wauwiler Moos, the point of which was hardened in the fire like the double-bladed paddle spades from south-east Africa mentioned above. In Pfyn-Breitenloo a paddle-shaped spade belonging to the Michelsberg culture was discovered (Müller-Beck, 1965, Fig. 124).

New evidence from New Guinea has radically expanded our knowledge of the range of tools once used in agriculture there. Some years ago Allen (1970) published a tanged and shouldered "hoe" blade of slate found at Kuk during initial drainage (Fig. 75). The blade is 17·7 cm wide and 26 cm long including the tang. The tang itself is about 5 cm wide and 7 cm long (cf. Allen, 1972). Another tanged blade of slate (Fig. 76, upper) was found at Wurup in the Manim valley south of Mount Hagen (Christensen, 1975b). This blade is about 31 cm wide and 27 cm long but only about 1·5 cm thick; the tang is about 12 cm long, 5·5 cm wide and 1 cm thick. One surface of the blade is slightly convex, the other flat. The drawing on the upper right of Fig. 76, shows the convex side, which was probably the front, that is away from the holder. According to Christensen it had been slightly smoothed or polished. But striae on both faces, in the depressions and on the ridges ran in all directions so that, in Christensen's opinion, they could not be due to wear from use. What was not noted on this artifact, however, is that between the base of the tang and the blade there are distinctive facets on both sides. Obviously these notches had been intentionally made and could have served as grooves to carry flexible circular ropes.

The lower drawing in Fig. 76 shows an artifact very similar to the Wurup piece above. This artifact was found at Mishrifé Qaṭna near Homs in Syria (Steensberg, 1964, 1977a). It was a tool that had been pulled by ropes, the wear of which can clearly be observed on both of the shaped shoulders at the transition from shoulder to tang. I have examined 13 specimens of this kind from Mishrifé Qaṭna and 15 from Hama. All dated specimens are from the Bronze Age, between about 2400 and 1800 B.C. Experiments with an exact copy of one of these artifacts showed that they were operated like an ard-plough, not like the traction spades used in Arab countries. That is, the blade made a shallow groove or channel by being pulled continuously along, not by being pushed into the ground, pulled by ropes and moving the earth to the side.

The Syrian rope-traction ards were mounted with a long shaft holding a transverse piece of wood between two prongs, the shoulders of the stone blade being slotted into the transverse piece. The stone artifact from Wurup may have been mounted in the same way, though its raised shoulders rather indicate that there was only

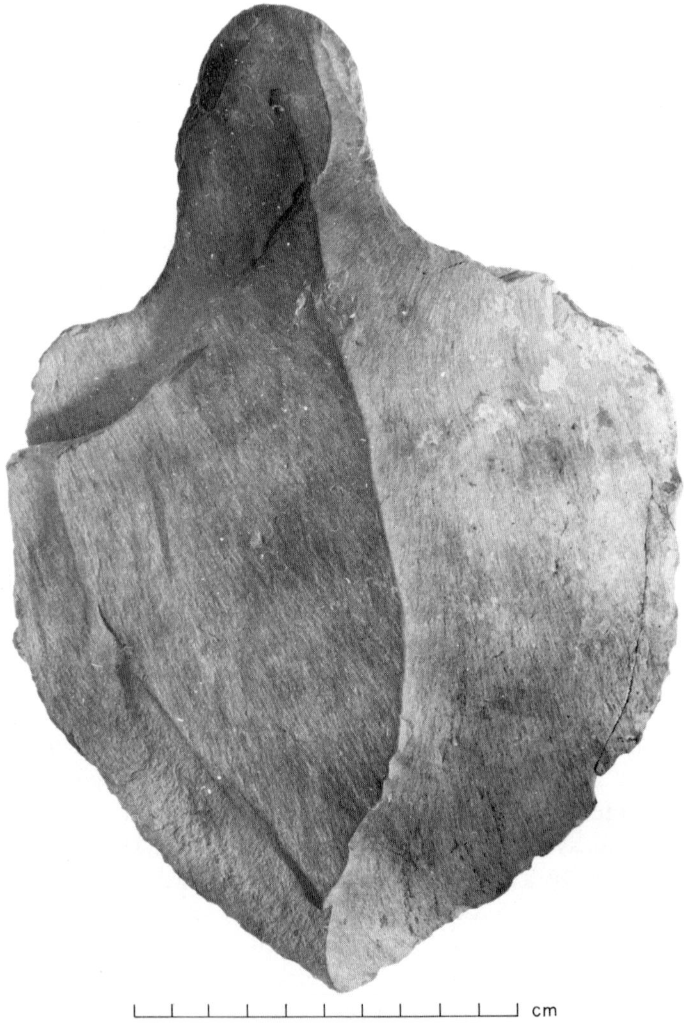

Fig. 75. A tanged and shouldered flaked "hoe" blade of slate from Kuk of unknown date. Reproduced with permission of Jim Allen.

a pronged shaft grasping the tang and fixed with a peg through the biconical perforation near the top of the tang. But while the Syrian tools were strong enough to penetrate 10 cm deep into a rather pebbly soil producing a furrow for the seed, the Wurup tool was thin and fragile. When I examined this artifact in the National Museum of Papua New Guinea in Port Moresby, I could not decide whether

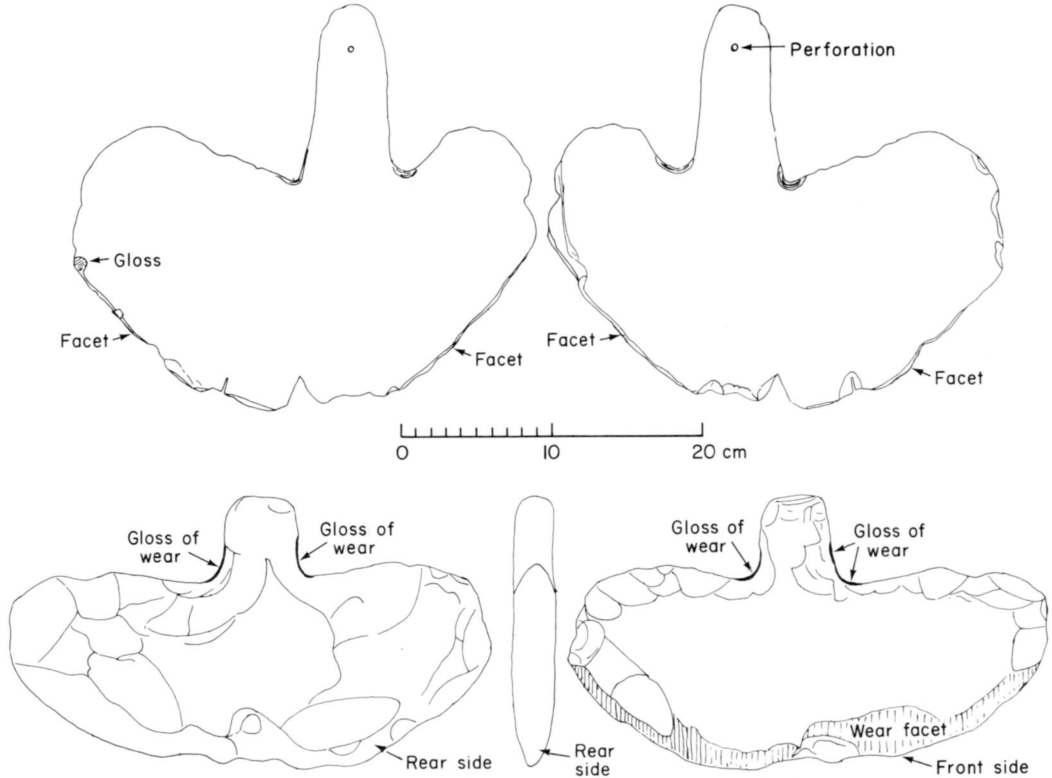

Fig. 76. Top, two faces of a slate blade from Wurup near Mount Hagen, date unknown. National Museum of Papua New Guinea, Port Moresby, No. E 14774, reproduced with the permission of the museum. *Below*, the faces and side view of a similar blade of hard limestone from Mishrifé Qaṭna in Syria, dated to the Bronze Age and interpreted as the share of a rope-traction ard for making furrows for sowing seed. Drawings of both by the author.

the grooves at the base of the tang carried wear marks like the Syrian specimens. Since nothing distinctive in the way of wear was noted by Christensen during his microscope examination, it is not likely that the piece had been heavily used, although its lower edge was damaged. This suggests that the Wurup tool might only have been used in very loose soil just like the rope-traction ard from Satrup Moor mentioned above; but it could hardly have been used for the sowing of seeds. It may have been a weeding tool, pulled between rows of plants, though it was more likely a tool for making small groove drains in gardens.

In a preliminary report Bulmer (1973) has referred to similar tools from her excavations in the Kaironk valley in the Madang Province. One of her illustrations (Fig. 5: Tanged Digging Tools from Wañlek and Kaironk) is reproduced here as Fig. 77, supplemented with a scale in centimetres in order to allow direct comparison with the Wurup and Mishrifé Qạtna tools in Fig. 76. Bulmer says that apart from the broken working edge the tool (d) is "otherwise the same size and shape as the 'hoe' collected by Allen from the Kuk site" and she continues:

> Some of the tanged tools from Wañlek are far heavier than would normally be interpreted as hand-held scrapers or knives.... As well, a number of the Wañlek tanged tools show evidence of irregular percussion breaks on their working edges [tool (a)], and many are snapped off across the working end as if by a heavy blow [tool (b)]. This use-wear contrasts both with the abrasive wear found on the "waisted blades" of the late Pleistocene Highland sites ... and the kind of wear on modern polished axes.... The tanged tools are made of schistose slate, which was broken into large flat slabs, requiring mainly edge-shaping and minor retouching ... some having a wide curved working edge, rather like a modern English grass-edging tool [tool (c)] ... all have a well-finished tang ... more carefully retouched than the rest of the tool ... smoothed following retouching ... they are presumably intended to be hafted in a wood handle.

Tool (c) is about 40 cm broad whereas the Syrian specimen mentioned above is only 32 cm broad, and the piece from Wurup 31 cm broad. But the blade from Wañlek is 2·5 cm thick, i.e. more than double the thickness of the Wurup specimen. We should be aware, however, when considering such implements, that use will have modified their shape. It is possible, for example, that the Wurup blade was more pointed originally, just as almost all the Syrian rope-traction ards must have been (Steensberg, 1964, 1977a).

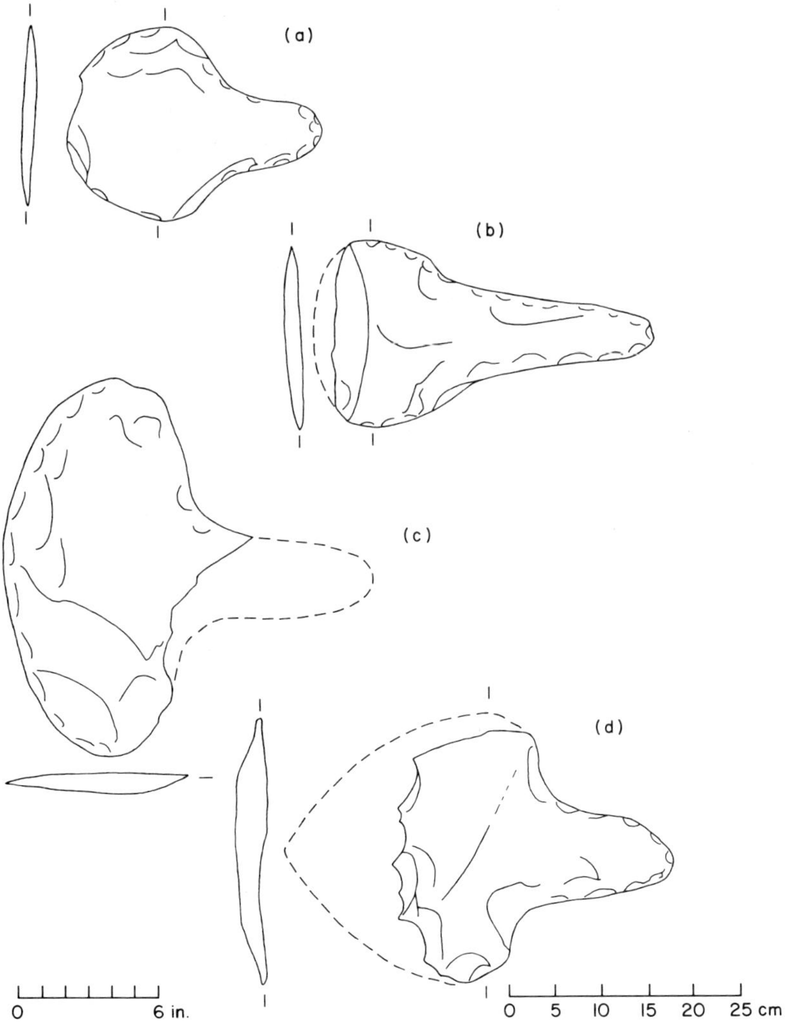

Fig. 77. Tanged digging tools from Wañlek in the Kaironk valley, Bismarck-Schrader Range, Madang Province. After Bulmer (1973, Fig. 5), with permission.

As far as the tanged blades are concerned, a connection between New Guinea and Asia can be suggested, because such blades have a wide distribution from eastern India to Hong Kong and the Yangtse valley in East Asia, as Wheeler has observed (1959, p. 87). But they also occurred from Japan and the Philippines to Malaysia, some of stone, others of metal. Some of the stone ones have grooved shoulders as can be seen in a number of illustrations in Duff (1970), one of which (Duff, 1970, p. 123) is reproduced here as Fig. 78. In the

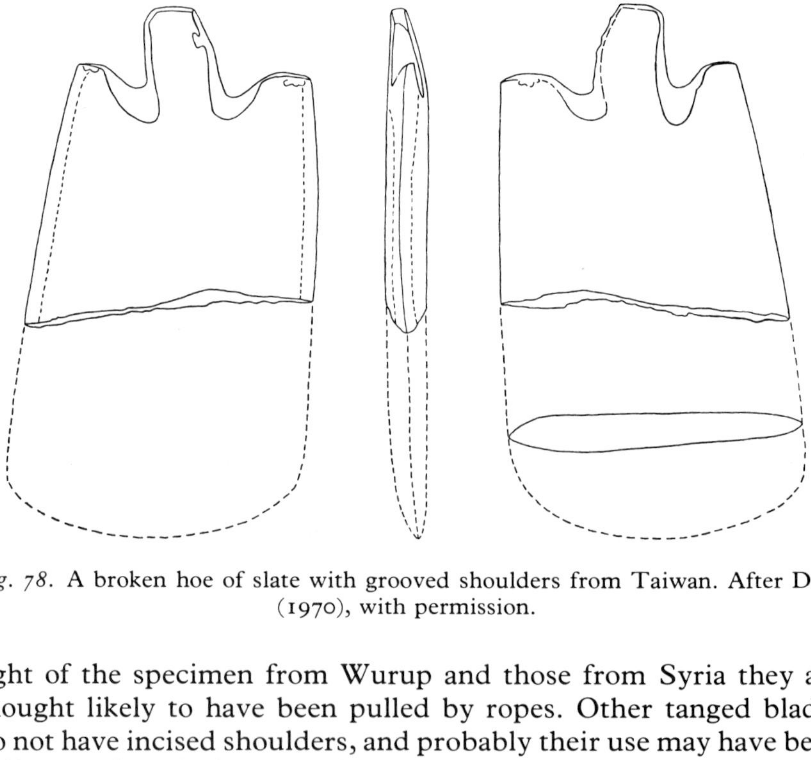

Fig. 78. A broken hoe of slate with grooved shoulders from Taiwan. After Duff (1970), with permission.

light of the specimen from Wurup and those from Syria they are thought likely to have been pulled by ropes. Other tanged blades do not have incised shoulders, and probably their use may have been different, though the main form is almost identical. Unfortunately all such tools have until now been treated typologically without taking possible wear marks into consideration, which could contribute to solving the problem of use.

The interpretation offered here of the Wurup blade as a traction implement is a novel one in the New Guinea context. Indeed, New Guinea has been regarded by the theoreticians of agricultural evolution as a place where even the hoe was not known and its agricultural technology has been classified at the lower level of the presumed evolutionary sequence. We must remember, however, that evolution is not unilineal, that alternative methods can be devised to solve similar problems in different environments and that, as far as New Guinea is concerned, many changes in the tool kit could have taken place over the 9000 years now claimed for the existence of agriculture there. Hoes have indeed been recorded in use in a few places in New Guinea in recent times, though not for tractioning. Allen (1970) refers to a shell hoe from the lower Fly river area (F. E. Williams' Anthropology Report, No. 14, Port Moresby, 1933), as well as to a wooden hoe hafted in the same manner as an adze, mentioned by Nilles from the south-east slopes of the Bismarck mountains (Nilles, 1942–1945). Nilles found that the Chimbu people of the upper valley did not use large paddle-shaped trenching spades because the slopes did not need to be drained artificially. This is like the situation in Enga where mounds characterize steep slopes and higher altitudes. The same explanation is valid concerning the people around Puyakira'go; they do not know of paddle spades because the land they cultivate is in rocky and steep country and their agriculture is less intensive, since their subsistence is derived also from hunting and collecting in the forest. However, in Alipe the soil was well suited to the use of paddle spades and also the high altitude made cultivation in mounds necessary. The Chimbu of the lower valley not only used the paddle spade; they also needed the checkerboard pattern of square beds separated by trenches because the ground was more damp. For weeding they made use of an adze-like hoe with a hardwood blade, about 40 cm long, 10–15 cm wide and 2 cm or more thick, with one side flattened and the other rounded. Nilles states that the flat side faced the flat "upper part of the elbow shaped haft" to which it was bound with strips of bamboo, rattan or the dry ribs of banana leaves, i.e. the flat side of the hoe blade faced the woman or man who used it. He explained the hoe in terms of an adze, because the hafting is similar, although the adze is used standing up whilst the hoe is used bending down, for crumbling compact clods turned over by the digging sticks. Nilles continues:

> Whether the hoe is occasionally used for breaking up the ground I could not tell. Its main use is to loosen the ground which had been cut for the trenches or to shapen [*sic*] and flatten the trenches themselves, and also as a weeding tool. As the use of the hoe is rather limited, its role being substituted by the digging stick, spade and even hands, it cannot be classified among the most outstanding agricultural tools in this 'trenching culture'.

It is worth considering whether the function of this Chimbu wooden hoe—working in loose soil, breaking up clods of earth, weeding—was not in the past performed by some of the tanged slate blades found in archaeological sites at Wañlek and as stray finds elsewhere in the Highlands, like the Kuk specimen (cf. Golson, 1977a, p. 622). They would have been suitable for such work, whether hafted like the wooden hoe described by Nilles or used in the hand with the forearm as an organic helve. It is interesting to record that an exact reconstruction of the Nilles hoe was made by the Museum für Völkerkunde in Leipzig and is illustrated by Damm (1955, p. 116, Fig. 30). Its angled shaft is 1 m long and the wooden blade attached adze-fashion to it is 40 cm long.

The nearest European parallels to Nilles' hoes are the shoe-last axes (*Schuleistenkeilen*) of the Danubian culture. They have a slender form similar to the wooden hoes observed by Nilles. It has been debated for many years whether they were really axes or were plough-shares adapted to the loess soils of Central Europe. Hennig, by considering the wear marks on them, seems to have proved convincingly that they could not possibly have been used for ploughing (Hennig, 1961). He argues that they must have been used for woodworking, like adzes of the same size and shape known from the Pacific. In my opinion some of them could have been used as hoes. The hoe hafted with an elbow handle is a weeding tool rather than a tool for breaking the ground, though it may be used for the surface cultivation of garden beds and scooping soil up around plants, as in Africa.

The hoe may have originated independently in different parts of the world and at different times, but in the main development of agricultural tools it seems to be of later origin than the heavy digging stick and the paddle spade.

3 Fencing Against Pigs

Pigs may have been in New Guinea for up to 10,000 years. Bulmer (1975, p. 18) has reported a pig incisor among the faunal remains of a Highlands rockshelter in a level dated to this period. Golson and Hughes (1976) have suggested that pigs were present at the Kuk archaeological site 9000 years ago. They are of the opinion that pigs, not being native to New Guinea, are most likely to have been introduced by man because the minimum distance of 95 km from the nearest Indonesian island at the time of the lowest Pleistocene sea-level could hardly have been overcome independently by a sufficient number of pigs to establish viable populations in New Guinea.

It is tempting to suggest that in this part of the world the breeding of pigs and the cultivation of plants were mutually related from the very beginning, because man and pig, having a similar system of digestion, shared the same kind of food. Their relationship depended on the pig's interest in man's food and man's interest in having a living store of flesh within reach, which could be transformed into meat anytime he needed it.

Probably feral pigs were hunted with bow and arrow, and in some places in New Guinea domesticated pigs are still shot with arrows on ceremonial occasions (Wirz, 1927, p. 86: Heider, 1970, Pls 11–12). As time went on and agriculture began to transform the environment through forest clearance, pigs became more dependent on cultivated foods (cf. Morren, 1977). The agricultural techniques to provide for this were available or developed, as were the means for controlling pigs in an agricultural context. Fences have probably always been necessary, not so much to keep the pigs in, but rather to keep them out. In the village the pigs are often tethered to a pole, but generally they are led out of their sties through fenced alleys into the bush where they can forage for most of the day. Clarke recorded the plans of two women's houses and one family house of the Angoiang in the Ndwimba basin showing such alleys leading from the pigsties (Clarke, 1971, p. 105, Fig. 5). Figure 79 is a view of an alley on the north side of the Kum river near Mount Hagen town. The pigs, stied in an abandoned women's house, are led between the fences of tilled gardens into the untilled swamps along the river, crossing deep drains on bridges consisting only of two or three long logs. Bananas are grown on both sides of the alley, providing tempting refreshment en route.

This particular fence consists only of vertical stakes pushed into the muddy ground, but in most places the stakes are tied to a horizontal stringer made of pitpit (Fig. 80). In other cases some of the stakes are tied together and others are loose, as I observed alongside a recently built settler's house at Kindeng on the Wahgi river. If the stakes themselves consist of pitpit, as in Alipe on the Kaugel river (Fig. 81), they are normally secured with a second stringer near the bottom. In the background of the alley shown in Fig. 81 there is a "gate" made of heavy, flat wooden stakes which are only tied together near the top. There is a gap in the middle of the gate, which is crossed by a twisted fibre rope, one end of which is tied to the middle of the first stake to the right of the gap, the other to the third stake to its left. The opening can be closed by pushing a few extra stakes through the twisted rope from below.

In another place near Mount Hagen I saw very tight fences of bamboo stakes, secured by a stringer formed by bundles of pitpit lashed near the top of the stakes (Fig. 82). At this particular place, near Wurerp Catholic Mission, a new road had been cut deep through the old surface and in the vertical wall old drainage ditches were visible, similar to those excavated and mapped in the swamp at Kuk. In some other areas fencing was sometimes only of planted *Cordyline*. Gardens on swamp land, as around the upper Wahgi river, were enclosed by a boundary ditch, perhaps 1 m wide and 1 to 1·5 m deep, the sods from which were piled up on the garden side as a low wall (Powell *et al.*, 1975, Fig. 3, upper left). The technique of digging deep trenches with wooden spades has already been described.

Where there is easy access to wood, fences are normally made of horizontal logs piled up between two rows of vertical stakes as seen around the village of Mendim near Kindeng (Fig. 83). Here the newly-built road is barred by a gate consisting of loose timbers, slotted in between two uprights at each end. If a vehicle comes along, the driver has to stop and lift the timbers away. Pedestrians can climb over the fence by way of the two oblique logs to the right of the gate, a common type of stile in this country. Another common stile consists of horizontal logs piled up on each side of a fence as in Fig. 84, photographed at Alipe in the Kaugel valley. Or, as in an example at Puya-kira'go, the stakes are shortened over a short length of fenceline and a single log provides a convenient step over them.

The stakes are often pointed at the top as in Fig. 84 and as illustrated by Clarke (1971, p. 161) who shows such a garden fence belonging to the Bomagai-Angoiang people of the Ndwimba valley. Heider (1970, p. 268) observed that the Dugum Dani compound in the Baliem valley had two fences, one enclosing the yard and the other the banana gardens behind the houses. Fences in this area would be constructed in two ways; either, like the walls of a house, with horizontal planks held together by pairs of uprights lashed together through the slots, or, as in the Hagen region, "with upright palings held together by an upper and lower pair of horizontal stringers". They usually have a thatch of grass "sometimes simply laid on top, sometimes supported as a narrow rooflet by short cross-bars. The thatching on fence tops are convenient places for laying things: drying plant stems and brine-soaked banana and so forth".

According to Le Roux (1948, p. 286) a type of fencing constructed by the people of central Irian Jaya consisted of pointed uprights, 1–1·25 m high, stuck into the soil at 0·25–0·5 m intervals in a single or double row. Long horizontal timbers were tied to them by means of rattan, a hand-breadth's space between them. These uprights were supported obliquely by props, mainly on the outside, but at greater intervals on the inside too. On top of this fence was placed a layer of palm leaves, weeds and grass which, according to Le Roux, was a protection against decay, though the pointed upper ends of the uprights protruded above the thatching material. Apparently the thatch kept the body of the fence dry, and rain water ran straight off the pointed uprights. An analogy may be drawn with the tops of European telegraph poles, which were formerly often sharpened like a roof and protected by a metal plate. For crossing such a fence there were a number of stiles which had higher uprights as handrails (Le Roux, 1948, p. 287).

I saw a fence built almost in the way described by Le Roux in

Fig. 79. A pig alley, formed between rows of vertical stakes, leading from the sties into the grassland. Kum river, near Mount Hagen, 1968.

Fig. 80. A fence of uprights tied at the top to a stringer of pitpit. Kindeng, near Mount Hagen, 1968.

Fig. 81. A pitpit fence tied to two stringers. Alipe, 1968.

Fig. 82. Below a bamboo fence a new road cutting reveals old garden ditches. Wurerp Catholic Mission, near Mount Hagen, 1968.

Fig. 83. A fence made of longitudinal logs with a stile of oblique logs and a gate of loose horizontal timbers. Mendim, near Mount Hagen, 1968.

Fig. 84. A very common type of stile made of a pile of short logs laid on the ground and held in position by pegs. Alipe, 1968.

the rainforest behind Port Moresby in 1968 (Fig. 85). However, the stakes were thinner and not pointed on top. Perhaps the openings between the horizontal timbers were to prevent standing water from producing rapid decay in the fencing wood. This was a swidden gardened for only two years, and because there was no village inside, the fence timbers would frequently be moved from one place to another. The fence was not protected by any thatch as it probably would have been if there were houses inside and the site used for a longer time. In Fig. 86 the swidden is seen from another angle showing the backdrop of dense rainforest.

Close to this swidden was a very small fence constructed around 3–4 taro plants (Fig. 87). They belonged to a small team of hunters, who lived nearby but did not possess substantially constructed houses. This fence is simple, with only one row of uprights, like those Le Roux mentions for Irian Jaya, the reason being that the garden had already been abandoned. The main fence had been moved elsewhere, but it was necessary to protect those few taro plants against feral pigs.

Le Roux also described another type of fencing consisting simply of slender stems from the swidden piled up along the inside of a row of uprights (Le Roux, 1948, p. 287). Such fences could be up to 10–20 m wide at their base. They remind one of the ancient Scandinavian *bråtegjerder* which consisted only of logs and branches, piled up as a very broad protective wall around a swidden. However, in olden times such a construction could also be an impediment to invading soldiers, especially if placed across narrow passes.

At Puya-kira'go the fences around the old site were constructed using a number of these techniques, as shown in Fig. 88. To the left is a fence of horizontal timbers between two rows of uprights but without thatching, and to the right a fence consisting only of vertical posts pointed at the bottom and embedded in the rather wet and deep surface soil at this particular place.

When I arrived at Puya-kira'go, my cook made a small enclosure behind my house where I could take my morning bath. He simply took some pointed planks and rammed them with all his strength into the moist soil with a twist very much like that used in the Mount Hagen region when opening up the grassland soils with the long paddle spade. Sometimes it was necessary to repeat the operation for a post to stand firmly in the ground. But another day I noted that the men used a club made of a natural knot in order to sink a post into the earth, just as people did in my childhood in Denmark.

Fig. 85. A horizontal fence in the rain forest behind Port Moresby, 1968.

Fig. 86. A further view of the clearing in Fig. 85.

Fig. 87. A fenced plot with some taro plants left in the regenerating forest after the recent abandonment of a swidden. Rainforest behind Port Moresby, 1968.

Fig. 88. An old fence consisting of uprights embedded in deep soft soil (foreground) and of horizontal logs between widely spaced uprights on the thinner soils of the rocky hillside behind Puya-kira'go, 1975.

The club was similar to those kept in the Danish Agricultural Museum at Gammel Estrup Manor under the local term *nødde*.

Simple problems can be solved by the same means everywhere in the world, but when the function of fencing is analysed, the local environment must be taken into consideration; not only the material available but also the consistency and character of the soil. The difference in the two parts of the fence in Fig. 88 is a consequence of the different character of the soil. To the right it was wet and deep because the surface formed a shallow basin, but to the left the soil was dry and thin and the limestone rock was not far below the surface.

In some places no fencing was necessary because the slope was too steep for pigs to climb. In other places a couple of heavy logs or planks were simply placed above each other along a hillside, sloping outwards and kept in place by slanting pegs (Fig. 89). Such fences would of course not last long. At the time the photograph was taken they were rather rotted and decaying, the pigsties not having been in use for some 3–4 years and the gardens abandoned.

In the Bismarck Mountains the principal gardens had to be placed on steep slopes (Aufenanger and Höltker, 1940, p. 38); gardens were sometimes situated an hour's walk away from the compound, and at least half an hour's walk was very common. Like the Karuka people around Puya-kira'go the Gende of the Bismarcks preferred to build their fences along a crest in order to save fencing material, the posts normally leaning out over the slope. In other places naturally forked timber uprights were placed 80–100 cm apart and a ridge pole put on the top; other posts, pointed at the upper end and anchored in the ground at the lower end, leant against this ridge pole and were tied firmly to it. This meant the points projected above the ridge pole towards the inside. Another row of oblique posts was placed along the inside, and the space between the uprights and the two rows of leaning posts was filled with logs. Such fences could be 1·3 m high and often *Cordyline* and cane grasses were planted on one or both sides. There were no gates, but stiles were made similar to that described from Mendim near Kindeng above.

In the Swiss Lake Dwelling at Egolzwill a fence was found on the landward side that consisted of uprights, the upper ends of which were also naturally forked, and in this fork a pole was placed. Apparently this was to protect the field against the cattle passing to and from the common. In the early Bronze Age settlement of Baldegg the buildings were surrounded by a single row of thin round posts placed so close to each other that no animal could pass through

Fig. 89. On hillsides, horizontal logs resting against outward sloping stakes are an effective barrier to pigs climbing the slope. Puya-kira'go, 1975.

them (Vogt, 1954). But fencing around settlements was also rather common in the Iron Age in Jutland. An elaborate system dated to the last 200 years B.C. was excavated at Hodde in west Jutland, Fig. 90 (Hvass, 1975, p. 80). An area 160 × 90 m was enclosed by an upright palisade of broad heavy planks positioned in a 40 cm wide ditch dug between 30 and 100 cm deep into the subsoil. The gaps between the uprights seem to have been about 15 cm, which indicates that they may have been interwoven with withes producing a tight and very strong fence. In fact this technique was the same as the Iron Age people of western Jutland used when constructing the wall

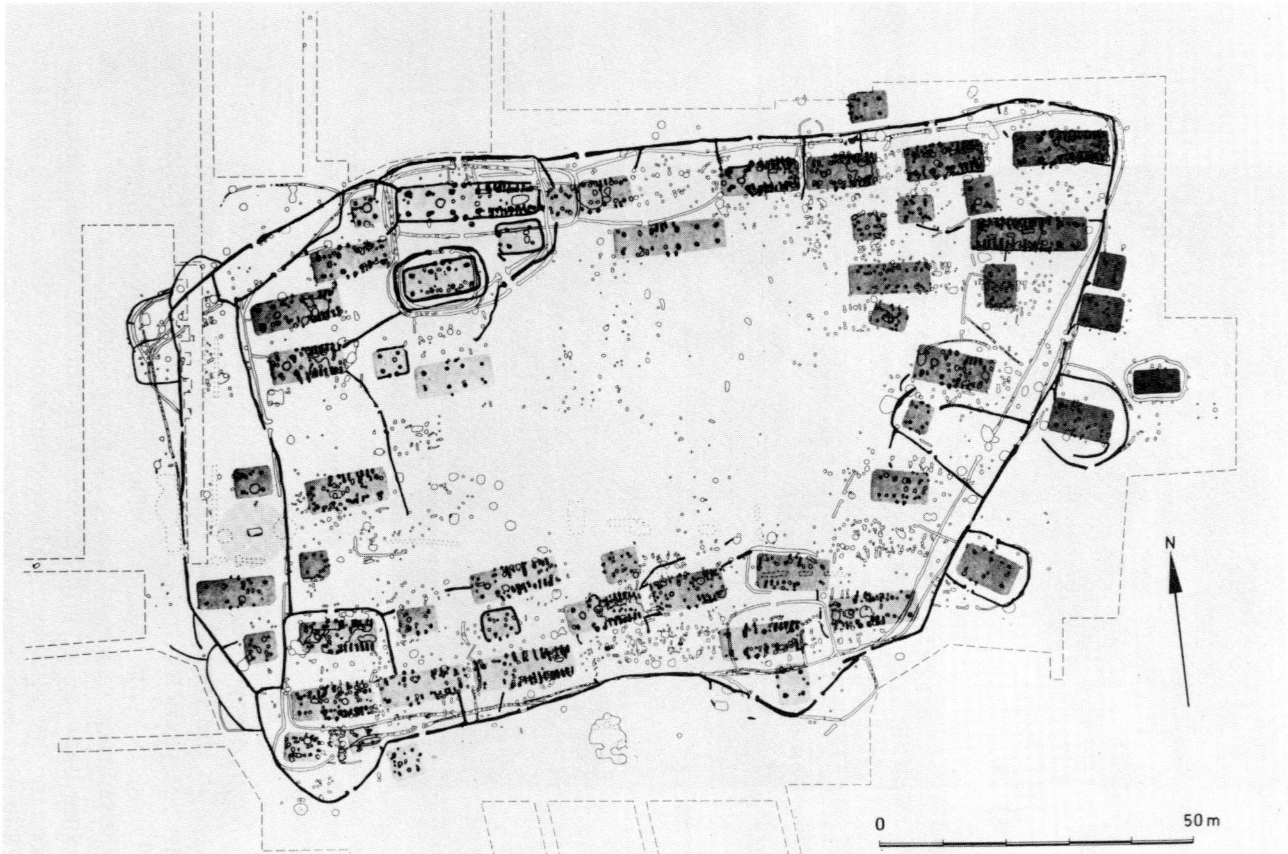

Fig. 90. Fenced village of the Early Iron Age excavated at Hodde, west Jutland. After Hvass (1975, p. 80), with permission.

of a house, though in this case the structure was plastered with mud. The palisade at Hodde enclosed a village of at most 22 farms. However, every farm inside the palisade seems to have been enclosed separately by its own fence, though not all the trenches for the fence were dug into the subsoil. Every farm had two entrances through its fence, one facing the centre of the village green and the other connecting it with the fields outside.

Comparing the purpose of this enclosure with that of the Bomagai-Angoiang mentioned previously, we find that in the Ndwimba valley the fencing was to protect a man's sweet potato gardens from the pigs, whereas in Hodde the larger enclosure must have been for the cattle on summer nights while the smaller enclosures were for protection against damage to the houses by cows eating the thatch and destroying the fragile plastered walls with their horns. However, the reason why each farm had an outlet through the main palisade is obscure. The excavator's comparison of Hodde with a medieval village with an open area, a "village green", where the cattle were kept at night during summer, will not explain this feature. According to medieval village law nobody was allowed to drive his cattle separately out into the hedged fields through a backdoor in his own farm yard. The cattle had to be collected on the village green and driven by the herdsman along the main drive into the common outside the hedged fields. Obviously the researcher was not aware of the difference between the medieval practice of open fields and common herding and the Iron Age way of farming with separate fields. At Hodde the cattle must have been herded separately from each farm, but how this was done in an area of "Celtic fields", arranged like a checkerboard, is a point to consider. Probably the cattle were tethered as pigs are in New Guinea.

As a conclusion to this chapter a classification and distribution map of fencing in New Guinea might be considered useful. However, even a tentative attempt would not be helpful, partly because of our present lack of evidence and partly because mapping could never reveal how intimately fencing methods are related to the geographical micro-environment. Many maps of such material features are misleading, assuming a distribution pattern the reasons for which are the cultural relationship between different regions. But as the example from Puya-kira'go shows (Fig. 88), local conditions may dictate different types of fencing within a distance of a few metres. If the visiting scholar does not examine the nature of the subsoil, his conclusions may be completely wrong.

4 Housing for Men, Women and Pigs

The relationships between men, women and pigs in New Guinea are in many ways similar to those of ancient Europe. During the Iron Age in Europe cattle were stalled at one end of the main house. As recently as 200 years ago west Jutland farmers used to keep a sow in a sty behind the kitchen, especially cared for by the housewife, and smallholders might have a cow or a calf in the living room itself. In New Guinea it is common for pigs to be housed indoors and for them to be looked after by the women. However, there is a striking difference compared to the European custom: in New Guinea men and women rarely live under the same roof and consequently the pigs are housed in the women's separate quarters.

The relationship between the inhabitants of any settlement and the housing that protects them is an intimate one, and it would therefore be unreasonable to discuss the practice of domestic husbandry and yet neglect the building-frames that encompass the activities of the inhabitants, like talking about a snail without its shell or a bird without its nest. A house is a reflection of three factors: the materials available in the environment; the activities undertaken by its inhabitants; and the size of the family and its social status, measured in terms of the volume of the house and the range of equipment and valuables owned. In all these respects it is useful to make a comparison between New Guinea and ancient Europe.

A proper analysis begins with the elements that support the building and tie it together: how the structure carries the heavy weight of roof and rafters, how walls and doors protect the inhabitants against wild animals and insects, wind and weather, or indeed against unfriendly members of their own species who might steal their stock and stores and possibly seek to take their lives. A simple round hut needs only one post in the middle, though it may have two or four, from which the rafters stretch down to the circular or oval wall. Houses built in trees in the New Guinea rainforest might mislead an old-fashioned evolutionist to place them at the base of his system as the "origin of species", a primitive abode in which man lived like an ape and from which all later house-types developed. However, a tree trunk, firmly rooted in the ground and supported with buttresses, is a more solid and durable centre post than poles dug into the earth, rapidly decaying without protection against soil organisms. And the construction of a house in a tree, moving in the wind, is no simple task. It requires great skill, as we shall see later in this chapter.

1. House Construction

It is no coincidence that we start with the structural elements of the dwellings instead of analysing their plans. Experience with the European evidence, spanning a continuous period of thousands of years from archaeological excavations and historical sources, shows that plans are flexible, readily adapted to different and changing needs, but structural elements are persistent and long-lived. In ancient farming societies, practice came before theory due to the limitations of the materials available and the local environmental conditions.

The features of house construction in New Guinea are similar to those in South East Asia, India and the Middle East as well as Europe. The basic elements are two or more uprights ending in a natural fork in which a ridge pole rests. This ridge pole carries most of the weight of the rafters running down to the walls on each side and of the purlins and thatch, as well, possibly, as that of the timbers laid longitudinally along the thatch to prevent the wind from tearing it away.

In Europe this type of construction can be traced back at least to the beginning of the Neolithic. The oldest Greek temples were built in this way, the naves on each side of a central row of posts having a statue at the same end, often a god and a goddess. Sometimes there were two rows of posts, producing three naves. At Knossos, two-naved and three-naved halls were found side by side. There was a central row of columns in the seventh century B.C. temple at Thermos in Aitolia and that of the sixth century B.C. in Paestum. In the seventh century B.C. the temple of Hera at Olympus had a three-nave construction with columns of oak (Robertson, 1945). In ancient Germania the forked column of wood was called *sul* (Lex Bajuwariorum: *Magensûl*), and this element survived in some regions of Central and Eastern Europe as well as in Scandinavia until the beginning of the twentieth century (Steensberg, 1977b, p. 91).

In New Guinea, as in early medieval Europe, wooden walls can be constructed either with upright staves or with horizontal planks, in the Nordic languages called *skiftesverk*, *lavegg* or *bul-construktion* (German: *Bohlenwand*). But because of the lack of metal axes, chisels, knives and saws in New Guinea, a refined construction with mortises and tenons comparable to the European way of joining timber was not developed. In New Guinea horizontal planks are still fixed between two rows of uprights, opposing uprights being tied together with bands of rattan, bamboo or other strong flexible material, as in the fences described earlier. All planks are cut out of tree-trunks in the same way as firewood is split with an axe, using one or two long sticks as wedges. Doors are normally made of loose horizontal boards, put up in such a way that only the inhabitants of the house know how to arrange them properly. But in interior dividing walls, there may be sliding doors guided above and below by runners and made out of a single large piece of wood, as in the man's house at Yuku mentioned below.

Above the fireplace there is usually a baffleboard or hanging shelf which is suspended from the rafters and deflects the sparks, flames and smoke rising from the hearth. This soon becomes covered with a layer of soot and though made of wood, it is not normally in danger of catching fire. In New Guinea, as in ancient farms in Europe, this shelf, in Danish called *røgloft* (smoke-loft), provides a depository for utensils which have to be kept dry, for fur and skin, and also for fruits and meat that are conserved by the smoke from the fire (Wirz, 1924, p. 97). Such a shelf was found in an Early Iron Age house at Mariesminde in north-west Jutland, charred from the fire that destroyed the whole building. As late as the first half of the seventeenth century baffleboards were often found in Zealand farmhouses and they were also common in north and western Germany at that time (Steensberg, 1974).

Chimneys are not provided in the New Guinea house, though in the Highlands the nights can be cool and the open fire is never extinguished. In Denmark, chimneys were introduced in farmhouses in some parts of Jutland in the seventeenth century, but they only became common after the middle of the eighteenth century. Until then an open fire on the clay floor was normal, though clay ovens for baking had been introduced in the Iron Age. In New Guinea baking is not a traditional practice because no kind of grain is cultivated, and so the open hearth persists. The smoke is caught by the rafters and thatch or escapes through cracks and crevices. Daylight finds its way through the same narrow openings and through the doorway, and at night the burning fire throws a fitful glare over the people squatting around it or stretched out on the floorboards at both sides, very much as in peasants' houses in northern Europe some centuries ago.

2. The Settlement of Yuku

Most of the settlements west of Kopiago and south of the Tumbudu river are dispersed in the bush. One may find a men's house and a women's house situated at a short distance from each other but separated from other settlements by secondary forest, scrub and land in fallow. In some cases a few neighbouring settlements might be termed a hamlet, but real villages do not exist except in the more densely populated Duna areas south-east of the Muller range.

Behind a tree-covered slope to the north-west of Puya-kira'go, there was one such lonely settlement called Yuku where Kualu lived with his wife Dayame. Here Kualu had levelled three small platforms in the hillside, loosening the soil with his digging stick and moving it with his bare hands. On one platform there were the ruins of Dayame's old house (shown on the map in Fig. 91). On another platform, above and to the east, there was an unwalled shelter for pigs, and on the highest platform, just below the steep path leading to Puya-kira'go, stood Kualu's house, and, to the west, Dayame's latest house with its night-time stall for the pigs. Her house was oriented east-west with the entrance on the east gable looking on to the rear of Kualu's house, which was positioned with its west end turned slightly towards the north. This platform sloped gently so that drains had been dug only along the sides facing the hill. However, logs were placed along all four sides of each house against the bottom of the walls to prevent rain water getting inside. The lower side of the platform was supported by more logs fastened to short poles driven into the soil. Two drains ran under the logs and

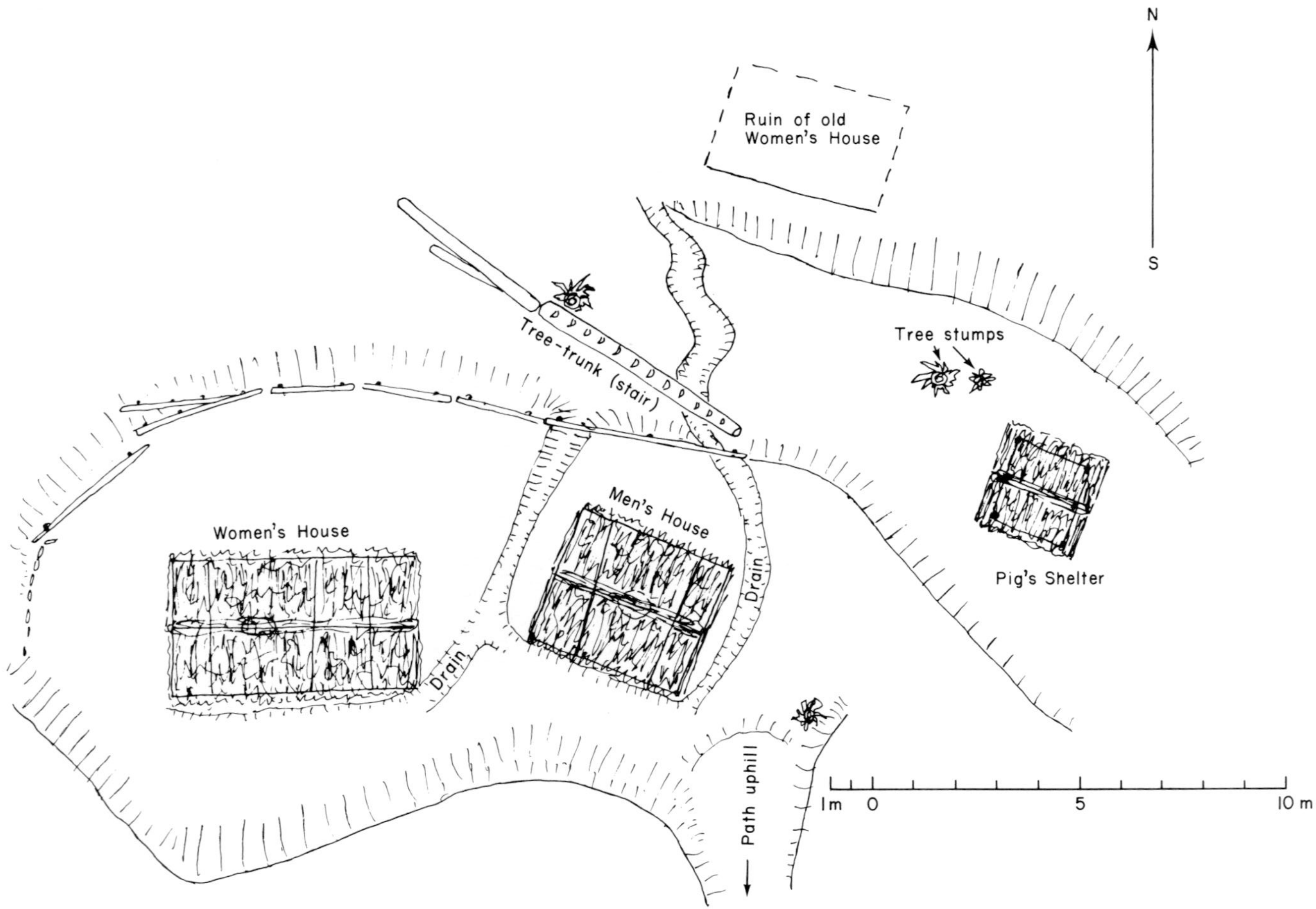

Fig. 91. Plan of the family settlement at Yuku near Puya-kira'go in the Southern Highlands Province, 1975.

the more easterly of these also ran below a heavy timber with steps hewn out with an axe, which functioned as a stairway from the platform down the steep slope to the gardens. About 100 ft to the east there was a fourth platform on which Kualu's farm-hand, a young unmarried man, had his small house. This was not measured, but it is shown in the foreground in Fig. 92.

Fig. 92. In the foreground is Kuala's farm-hand's house. Behind a new family house is being constructed by Kuala for him and his wife, Dayame. Yuku.

In Fig. 93 centre, there is a plan view of Kualu's house, his *anoa anda* (men's house). This was a "generation" older than Dayame's house and when I measured it in 1975 it was probably four or five years old, which is an unusual age for a house in this region. It had been moved from a site more to the east, where a new combined men's and women's house was then being built, and as a result of its age its condition was rather poor. The house had two porches, one at the front entrance, and one at the rear, the latter being used mainly for firewood. The front and rear walls were set at oblique angles to the side walls so that the room inside was 2·25 m along the north wall but only 1·75 m along the south wall. The

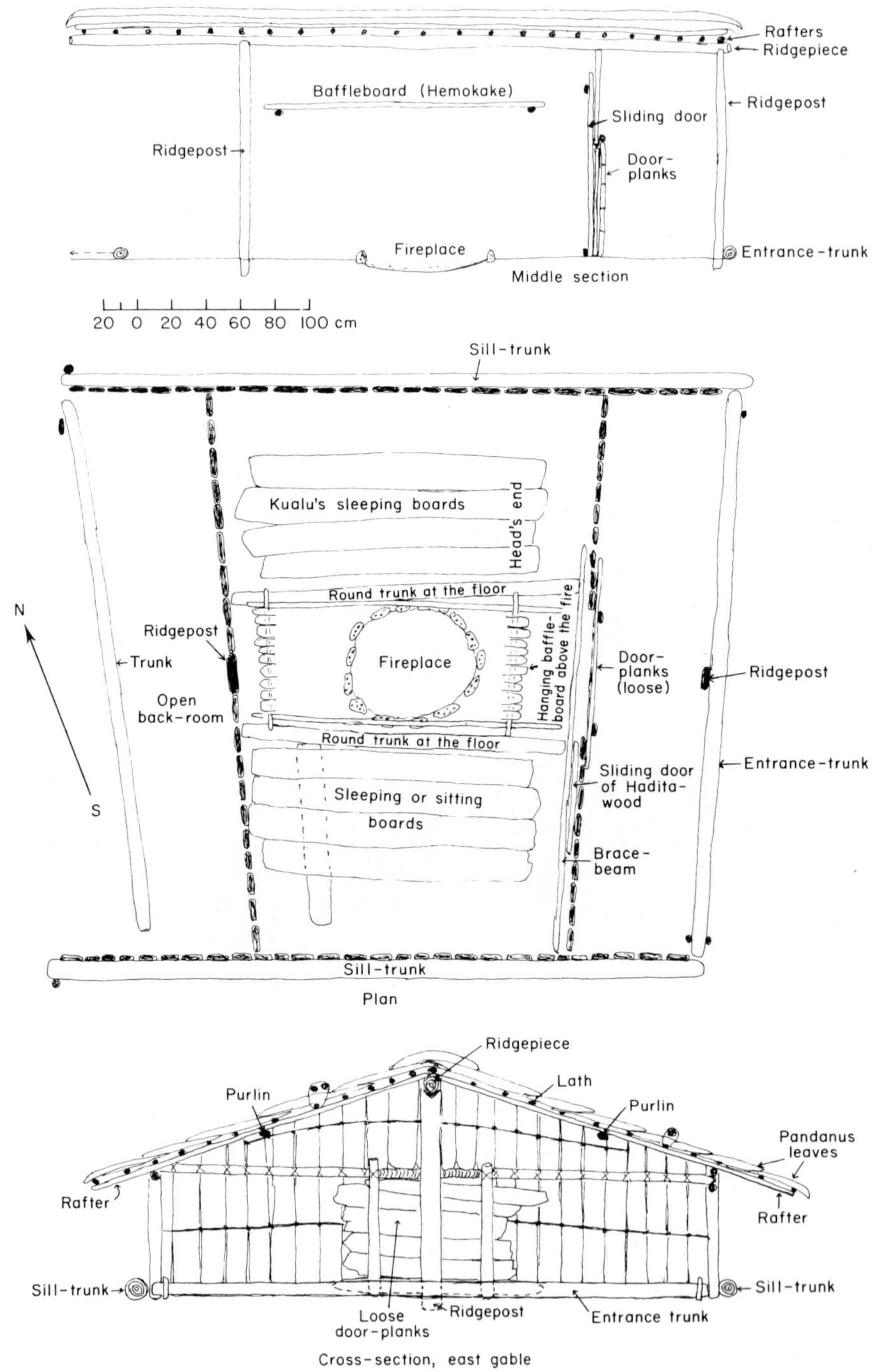

Fig. 93. Kualu's men's house, plan, elevation and section. Drawn by the author.

distance between the parallel walls was 3·35 m and the area of the room was 6·7m². The log along the rear gable had been displaced at the south end (Fig. 94) because three of the pegs which had supported it had broken or decayed. Also missing were one peg at the north end of the log, and two that had been at the eastern end of the two logs at the base of the two long walls of the house, which helped to keep the bottom of the wall staves in place.

An outer door consisted of horizontal boards slotted behind two uprights, the lower ends of which were driven into the soil and the upper ends tied to a stringer to which the wall staves were also fixed (Fig. 95). The inner door was a 70 cm broad and 110 cm high board cut out of a huge tree, sliding between runners above and below. It was only shut at night when Kualu was asleep. When it was locked and the planks of the outer door were set, it must have been difficult for anybody to break in without violence. In the women's house there was no such double protection. Usually in addition to Dayame, a young unmarried woman slept there with her two children.

From both sides of the doorway, two round logs placed on the earth floor stretched right over to the rear gable, and between these the fireplace was confined, the normal custom in this region. The fireplace was a shallow, almost circular hollow surrounded by upright stone slabs which acted as a kind of guard and kept the ashes and embers inside. Timbers for the fire were normally 0·5–0·75 m long, although sometimes they could be 1·5 m long, and the method of feeding the fire with them was either to push one end into the fire or to place them across the fire on top of the stone slabs and parallel to the two side trunks. As the timbers burned, the unburnt parts were gradually pushed into the fire until the whole log was consumed.

In Kualu's house the baffleboard protecting the roof from the fire hung 85 cm above the floor. Broad planks running parallel to the logs confining the hearth served as a sitting area, and those on the north side were used by Kweldu as his bed, his head facing the front porch. Over his bedhead hung a wooden comb with long teeth, and nearer to the wall was his net bag. In the corner there were several other utensils and his bow and arrows.

The walls consisted of upright wooden planks (Fig. 96), the lower ends of which were pointed and anchored in the soil as described earlier for fencing. At the top they were fixed to a horizontal beam or wall-plate on which the lower ends of the rafters rested. On the central line of the house, two ridgeposts supported a horizontal ridge-

Fig. 94. Kualu's house, a view of the west (rear) porch and gable.

Fig. 95. The entrance to Kualu's house barred by six loose boards

Fig. 96. The north side of Kualu's house consisting of upright planks lashed with rattan to a horizontal stringer. The round log at the base helps to keep the uprights in place.

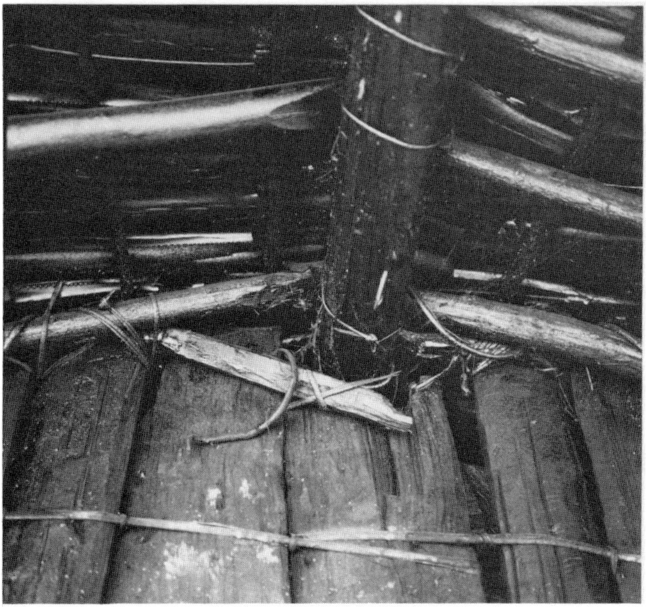

Fig. 97. The ridgepole inside Kualu's house, blackened with soot.

pole at a height of 1·25 m above the floor (Fig. 97). Two purlins, one at each side, were placed between the ridgepole and the wall-plates, resting on the posts of the front and rear walls. Rafters were laid across these purlins from the ridgepole to the wall-plates, the eaves overlapping the walls by about 20 cm (Fig. 94). A number of laths rested on the rafters, parallel to the longitudinal axis of the house. The lower 80 cm of the roof was covered with a layer of twigs and leaves, with the leaves overhanging to keep the rain water away from the walls. The upper part of the roof up to the ridge was covered

Fig. 98. Kualu's house seen from the slope above. Note the cleared area on the slope behind.

with the leaves of a palm-like tree called *kuaki* placed parallel to the laths. On top of this foundation layer the roof was "tiled" with overlapping pandanus leaves which were held in place by boards and poles (Fig. 98).

The very low roof and the presence of the baffleboard contributed to the cave-like atmosphere of the house: people could only move in a squatting or creeping position or with the upper part of the body bent forward. There was the advantage, however, that the heat from the fire was retained, for Yuku is about 1900 m above sea level and the nights are rather cold. As no bed-clothes were used in this area, a fire was kept burning or some embers glowing throughout the night. A severe disadvantage, on the other hand, was the smoky atmosphere. Respiratory diseases and shortage of breath were common in this region.

Cleary and Blackburn (1968) investigated air pollution in three huts in the Eastern Highlands and six in the Western Highlands of Papua New Guinea and there appeared to be a correlation between this and the high prevalence of chronic non-tubercular respiratory disease in the native population. As might be expected, there was more smoke immediately after lighting the fire in the early evening than at other times. Overall the inhabitants of the Eastern Highlands were more exposed to smoke pollution than those in the Western Highlands because the former slept on raised platforms approximately 18 in. (46 cm) above the floor.

Another report of an investigation into pollution was published by Master (1974). He also referred to the inhabitants' scant use of clothes which made them burn fires to keep warm during cold nights and he noted that extremely high levels of particulate matter and aldehydes were inhaled.

> Pulmonary disease, mainly obstructive but also restrictive, appears at an early age and was present in 78 per cent of subjects over age 40 years. Severely affected subjects have diminished breath sounds, coarse rales, and decreased chest cage movement. Pathologic specimens from affected subjects demonstrated centrilobular emphysema, thickened pleura, pulmonary fibrosis, mucous gland hypertrophy, and deposition of anthracotic pigment. Other factors probably responsible for the high prevalence of pulmonary disease include smoking of home grown tobacco, protein malnutrition, poor sanitation, and various diseases.

I have heard people coughing in the men's house at Puya-kira'go at daybreak, and such experiences lead an archaeologist to consider how air pollution may have troubled people in similar houses in Europe before the introduction of the stove and chimney.

The women's house (*ima-anda*) at Yuku was much longer than Kualu's because there also had to be space for the pigs. In fact the small shed mentioned above was originally built because there was little room for pigs in the former women's house, but at the time of my visit it only served as a place where the pigs liked to rest in the shade about midday, and they had made a wallow in it in order to cool off. The gables of the women's house were constructed like

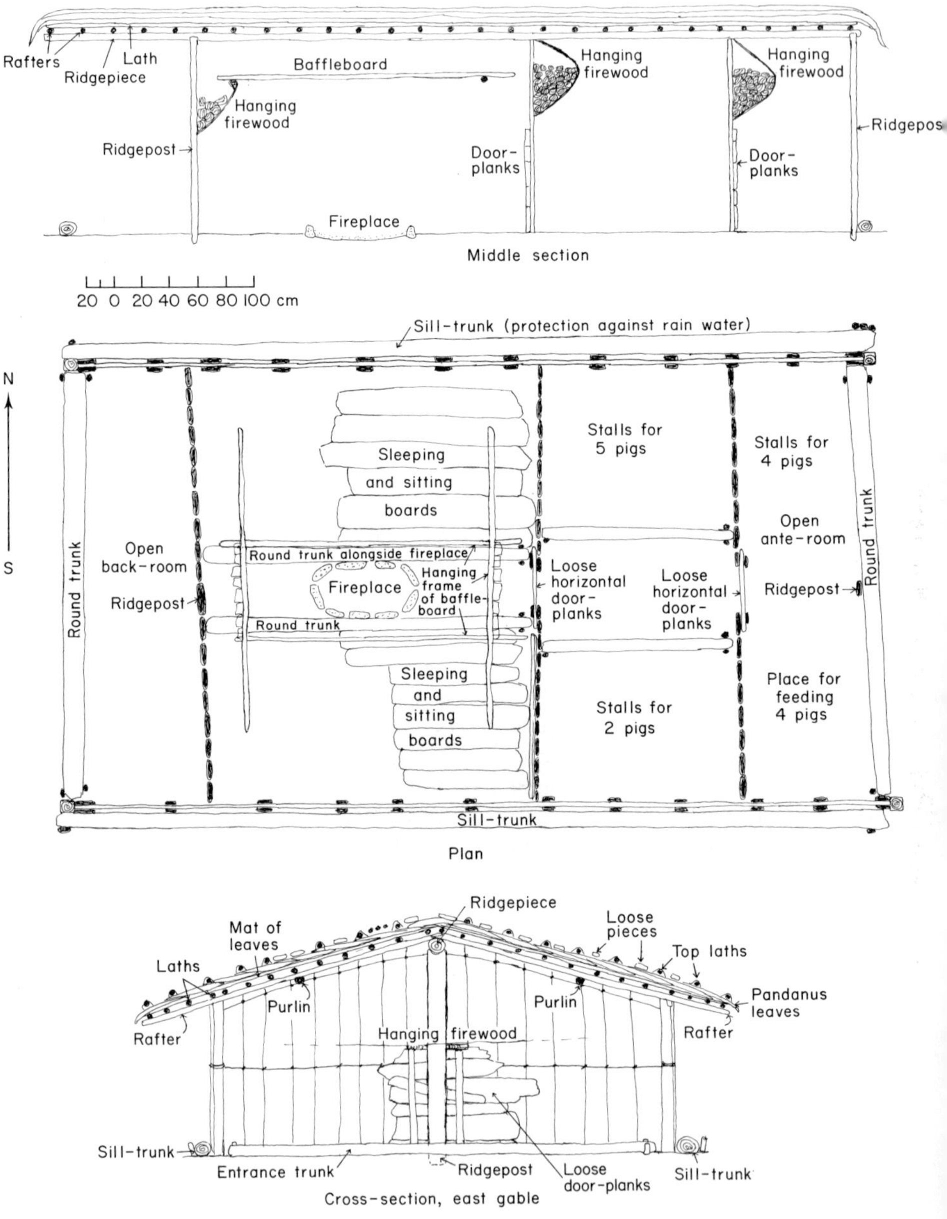

Fig. 99. Dayame's women's house at Yuku, plan, elevation and section. Drawn by the author.

those of the men's house (see Fig. 99), but the side walls were made of horizontal planks fixed between uprights 30–60 cm apart. Firewood hung below the eaves (Fig. 100), above the front door in the porch, above the door from the pigsties to the living room (Fig. 101) and on the inside of the gable.

Dayame used to sit in the porch and work at net-making and other tasks (Fig. 102) when she was not occupied in the garden. The place where she is shown sitting in Fig. 102 was used at night as a feeding area for four pigs, which slept in the opposite corner of the porch. Through the porch were the normal pigsties situated on both sides of the passage, formed between two round logs, and leading to the interior living room door. Dayame's living room was normally a forbidden area for men other than her husband, but as a foreign guest I was allowed to enter it in order to measure it and to photograph its fireplace, baffleboard and other features.

Through its entrance one could see the oval fireplace, surrounded by stone slabs as in the men's house, and the firewood hanging against the back wall (Fig. 103). The log to the right was out of its original position, having been displaced to the left, towards the back, because of the pressure of the mass of firewood. Figure 104 shows the north-west corner of the house with another bundle of firewood tied up along the north wall, and yet more in the corner. Note the empty milk-tin and net bag placed on mats made of pitpit.

Figure 105 is a photograph of the baffleboard seen from below. It was covered with a thick layer of soot from which dendritic growths hung down, gleaming in the light of the flash. In the background is the inside of the west gable with firewood hanging in rattan straps. Figure 106 is a view of the outside of the west gable with firewood hanging in the open porch. In front of the gable is the place where Kualu used to split trunks and thick branches into firewood (cf. Fig. 91). The roof of the women's house was similar to that of the men's house, but on top of the pandanus leaf layer there was a regular covering of sticks and boards to protect the roof against the wind.

The local indigenous Catholic priest had told Kualu that a Christian man should live under the same roof as his wife. Kualu had therefore decided to build a common house for himself and Dayame further to the east on a spot where his own man's house had stood before it was rebuilt. The plan of the new "joint" house is shown in Fig. 107, together with an elevation of the east gable and a longitudinal section through the median line. The fireplace

Fig. 100. The north side of Dayame's house. Firewood hangs against the wall, and at the base the long timber helps to keep the uprights in place and provides protection against the occasional flood; its original cover has partly eroded away.

Fig. 101. Inside Dayame's house firewood is stored in a sling above the door from the pigsties into the living room.

Fig. 102. This family shot shows Kualu in the foreground making a rope and Dayame sitting in her porch occupied with net making. A young girl visitor in a grass skirt is chatting with her. Note the rain canopy made of pandanus leaves on the roof.

Fig. 103. View into the living room of Dayame's house showing the hearth full of ash and the firewood stored behind.

Fig. 104. The north-west corner of Dayame's living room with stores of firewood and household utensils.

Fig. 105. The baffleboard in Dayame's house seen from below. Soot from the fire has produced dendritic growths that hang down from the flat canopy.

Fig. 106. The west (rear) gable of Dayame's house.

of the former men's house was still visible as a burned spot on the ground, and the new house was placed so that this old fireplace would function again in the centre of the new *anoa anda*, Kualu's own room. However, there would be no door between his and his wife's room. In this way old tradition was combined with Christianity adopted here within the last ten years or so.

Figure 108 shows the house under construction, seen from the east. The side walls were made of horizontal planks as in the women's house, slotted in between double uprights, and the two gables and the dividing wall between the two rooms consisted of upright planks as usual. The ridgeposts and ridgepole were also made as in the other two houses, as well as the rafters and laths. However, as this house was under construction, full advantage was taken of the opportunity to study thoroughly all the rattan bindings, as demonstrated in Figs 109–113 and explained in the captions. Because such details have not been studied elsewhere in the area, it is impossible at the moment to relate them to configurations of other areas and cultures. They may seem to be things of slight importance in themselves, but are crucial to an understanding of prehistoric structures, the vestiges of which are uncovered by archaeologists. Without them it will be

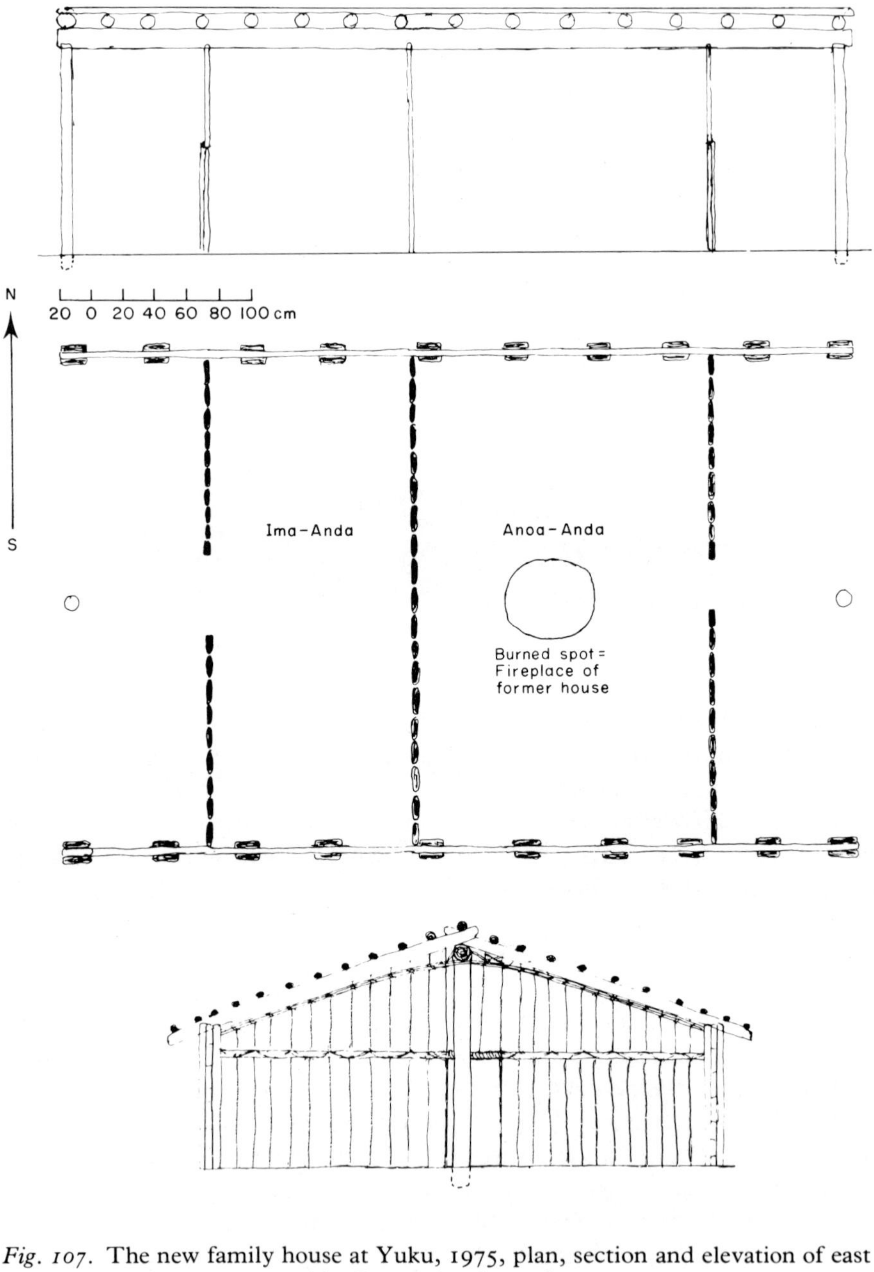

Fig. 107. The new family house at Yuku, 1975, plan, section and elevation of east gable. Drawn by the author.

Fig. 108. The new family house at Yuku seen from the east.

Fig. 109. Detail of lashings at the east gable of the new house. The ridgepost at the porch has not yet been put in place.

Fig. 110 (*left*) Lashings at the western gable of the new house. The ridgepost will be set in the ground in front of the porch to support the protruding ridge beam before the thatch is laid. *Fig. 111* (*right*) The south-east corner of the house showing the rattan bindings tying the different elements together. Except for the two at the eaves the laths were second-hand; hence their soot-blackened state.

Fig. 112 (*left*) The dividing wall between Kualu's and Dayame's rooms seen from the south. This picture shows how the split rattan was wound along the rafters, making a loop around every lath. *Fig. 113* (*right*) The western entrance to Dayame's room seen from the south. The uprights of the side wall and those of the inner dividing walls were hammered into the soil, as can be seen from their frayed tops.

impossible to provide the true details of traditional housing in any future open-air museum, displaying traditional Papua New Guinean culture of the type which has proved so popular and instructive in other parts of the world.

Most of the 28 laths and 26 rafters used in the joint house, examples of which are seen in Figs 111–113, were taken from the ruined women's house and were well smoked and protected against decay; only three laths and one rafter were new. Both the ridgepole and the purlins were re-used, and in some cases even the old wall staves had been incorporated, either in the same position or turned upside down and hammered into the ground. The new walls were a little lower than the old ones, because the ends of the old planks had rotted somewhat through being embedded in the ground and had to be trimmed off. Though the people are well aware of the effects of fire-hardening, which they use to make pointed tools more durable, they apparently do not treat the ends of planks and posts that are to be buried in the ground in this way.

Having used the best material from the former women's house in the construction of the new family house, the remainder would be soon removed for fencing or firewood (Fig. 114). Every last stick would probably be put to some use or other. Most of the old wall against the slope in Fig. 114 was still standing because the majority of its uprights had decayed so much that they could not be re-used for wall construction. On the terrace to the left the pig shed is visible and in the distance to the right Kualu's *anoa-anda*.

It is the men's work to build houses and they do it with great pleasure, in former times with a stone axe or adze as their only effective tool. Every three, four or at most five years they have to repeat the work, and in this region it is done with only one or two helpers, sometimes the older boys, sometimes a young man working as a farm-hand in the gardens and forest.

The rapid decay of timber structures in the tropics is one of the main reasons why houses and settlements are so readily moved in societies based on shifting cultivation, where the gardens have to regenerate by fallowing. Even in a comparatively stable settlement like Yuku, people have to rebuild their houses on a new spot at frequent intervals, though they do not normally renew both the men's and the women's houses in the same year. At Yuku the same platforms or terraces have been periodically re-used within a distance of about 50–100 m since Kuala was married 20–25 years ago. He told me that until the white man arrived in Kopiago in 1961, the

Fig. 114. The ruins of the former women's house at Yuku after the removal of most of its timber for the new house. The pig shelter is centre left and Kualu's house in the distance right.

local people had only stone axes and knives of bamboo for the building of houses. He estimated that with those tools it took four to six times as long to build a house as it does today with steel axes and knives.

We must contrast also the long and short-term stability of settlements. In the wild karst landscape around Puya-kira'go, a family normally has its own area confined by natural boundaries such as streams, valleys and defiles, and everyday movement of the family group is restricted to this area of forest and scrub, fallow and garden. In other areas with a more complex system of settlement, such as in the Ndwimba basin described by Clarke (1971, p. 100), everyday movement can be more extensive inside the same clan boundaries. In the longer term families can move their residence quite frequently. Of 86 Horaile men, interviewed by Modjeska, two-thirds had lived in more than one parish during their lifetime (White *et al.*, 1977, Table 8) and presumably a similar proportion of women did likewise. Even in the wilderness areas of the New Guinea Highlands, there will be few places left unclaimed by anyone—just as the immense forests of Finland are "owned" by somebody, and a somebody who knows precisely where his boundaries are.

3. The Men's Houses at Yaku and Puya-kira'go

At Yaku to the west of Yuku, there was a small catholic chapel behind which was a traditional men's house, as stocky as a fungus growth. Its rear gable faced the hillside, banana and pandanus had been planted decoratively around it, and there were wide views from the front across the valley to the distant mountains.

The roof was thatched with pandanus leaves as in Yuku and anchored with long thin timbers (Fig. 115). The side walls were made of slotted horizontal planks as in the women's house at Yuku. Thick timbers hung under the eaves, stored for the making of tools or as a reserve of firewood. A supply of firewood also hung from the upright staves of the gable wall in the rear porch. Round logs were placed along the foot of the walls and at the front of the porch, and the gaps below them were filled with mud so that the water would not flow in and wet the floor. Along the eaves and the front of the roof, protruding leaves and twigs spilled heavy rain from the roof straight on to the ground. Figure 116 shows a part of the side wall and to the left the front uprights of the porch. In the middle one can see how the rattan, binding the inside and the outside uprights together, was secured: the end was tucked behind the lashing, probably pushed down with a man's bone dagger or a pointed stick, and then turned upward and once again twisted behind the lashing, the free end finally pointing obliquely downwards.

A close-up of the entrance (Fig. 117) shows a bundle of firewood hanging from the gable over the doorway and there are more pieces piled up at either side of the door. A broad board leaning against the gable to the left of the opening was, however, the door itself, made like the sliding door in the men's house at Yuku. This seemed to be the only device for sealing the entrance of this particular house, though some of the short broad timbers to the left might also have been used as horizontal door planks on the outside. Also visible are some bows and arrows which had been put down while the owners prepared a cooking fire outside. The young man sitting on the front log of the porch was dressed in the traditional fashion of the region with leaves of cordyline (in pidgin, *as gras*) hanging from strings around his waist. In addition he had acquired a leather belt into which was tucked the ever useful dagger made from the leg-bone

Fig. 115. The men's house at Yaku near Puya-kira'go: view of the front porch.

Fig. 116. A close up of the split rattan bindings of the corner posts and one of the uprights at the entrance of the men's house at Yaku, seen from the outside. The end of the rattan strip was tucked behind the lashing and then turned upward and once again tucked behind the lashing of the upright to the right.

Fig. 117. A young man relaxes on the front log of the porch of the men's house at Yaku. Other men have put their bows and arrows against the bundle of firewood slung above the door.

Fig. 118. View of the inside of the men's house at Yaku. To the right is the fireplace, to the left the floor is planked for sitting and sleeping. Note the light coming through the chinks in the gable.

of a pig or a cassowary. Note also the grassy-leaved decoration in his hair.

The inside of the house consisted of a single room, floored on both sides of the central fire with boards running parallel to the gables (Fig. 118). The men used these boards to sit on while chatting round the fireside. The fire was protected as usual by a round log at each side and a circle of stone slabs, many of which were cracked and broken from the heat of the fire. The photograph shows clearly the numerous chinks in the stave-built rear gable, and when the front door was open there was always enough daylight inside for some work to be done.

The Catholic chapel at Yaku was thatched with bundles of grass using what seems to be a modern technique similar to that of European straw-thatched farmhouses (Fig. 119). The bundles were fixed to the laths by rattan bindings, as can be seen to the left in the photograph. The house at Puya-kira'go built as a guest-house was also thatched with bundles of grass. When I was there, young men were starting to repair its leaky roof. Bundles of pitpit, collected and made up by the women, were thrown by a boy to the thatcher

Fig. 119. Thatching the Catholic chapel at Yaku.

who merely pushed them into the existing thatch where the leaks seemed to be. A round pole leaning against the eaves served as a ladder to the roof.

Another house in the area was a men's house with a traditional fireplace and a porch in front. The walls were built of upright planks, but they were covered outside with mats woven in the same way as in the Hagen district which we shall describe below. This makes a tighter construction but unfortunately has the disadvantage of retaining smoke inside, increasing the risk of injury to the lungs. The old man living in the house (Fig. 120) constantly complained of chest pains and became breathless after only a few steps. Although he was hardly over 50 he could no longer work. This particular house was so old that it had subsided towards the right front side, and the owner had had to shore it up with oblique supporting poles. From the front door the house was divided into two rooms, the inner one being used as a sleeping room.

Behind this house was a pig shed (Fig. 121) the roof of which was steeper than usual for this region, being constructed using two high ridgeposts carrying the ridgepole. The walls were also made in a non-traditional fashion with upright poles carrying a wall-plate and then filled with vertical planks. The shed however was defunct, for pigs were no longer kept at Puya-kira'go, because there was no wife or gardening. All that remained were the old man and his friends who visited him in the old men's house at night to talk, laugh, often cough when the fire produced too much smoke, and roast sweet potatoes in the embers. Apparently the men enjoyed themselves, for they sat all evening until ten or eleven o'clock when they fell asleep to the sound of the cicadas and the pouring tropical rain.

4. Dwellings in Trees and Tent-Shaped Houses

Local people at Puya-kira'go told me that the Hewa people in the steep mountains north of Lake Kopiago lived in trees. The patrol officer, Peter Turner, dissuaded me from going on a trip to this region because the precipitous nature of the country required some mountaineering ability. Such an environment seems to be an important reason for people building in trees. In 1975 Thomas Marecek of the University of Arizona visited the Hewa area and photographed an elevated house built on the lopped-off tops of tree trunks (Fig. 122). The floor was about 10 m above the ground, so that the inhabitants, a large family of about 20 members, had to

Fig. 120. The men's house at Puya-kira'go, shored up by poles.

Fig. 121. The pig shed at Puya-kira'go. To the right a corner of the men's house, showing the matting walls.

Fig. 122. Elevated house built upon tree-trunks in the Hewa region north of Lake Kopiago. Photograph by Thomas M. Marecek, 1975. With permission.

enter by a rope ladder. Naturally the tree trunks would last longer than posts dug into the ground, but Marecek noted that long poles had been put up between the trunks and some of these had begun to decay after only two years. Thus the house would still have to be partly rebuilt within two or three years. Besides such very elevated houses there were also long houses built on piles similar to those of coastal regions.

Schultze Jena on his way to the upper Sepik made a sketch plan of a tree house at Krissi near the Bewani Mountains (1914, pl. VIII). Figure 123 is taken from his book and shows that the house was constructed in the usual way with a ridgepole supported by two posts, the lower ends of which were forked and fitted over timber running parallel to the ridge. This timber rested on two cross-bars which in their turn rested on two side timbers fixed to branches of the tree. The roof was supported by two wall-plates and two purlins. The rafters extended outside the walls, carrying a horizontal lath on each side supporting the eaves. The floor was built about 2 m above the basic support structure and consisted of two layers which provided additional insulation. The fireplace was sunk into the "sandwich" and rested on the lower layer about 20 cm below the floor proper. Schultze Jena described the construction of the hearth as follows: at the bottom there was a layer of ribs of sago leaves and above this a layer of various smaller leaves, which was followed by a thickish layer of earth on which the ashes of the hearth rested. These contained some stones which may have been placed originally around the edge of the hearth.

Many years ago, tree houses were also built in the Astrolabe Mountains behind Port Moresby. Figure 124 shows a hut in a Koiari village built 10–12 m up in a tall tree the branches of which had been pollarded (Lindt, 1887, p. 36, pl. XIV). The construction was very similar to that described by Schultze Jena, but this hut apparently had two full storeys of which the upper one was the more important. A long rope ladder made of rattan led from the ground to the upper storey and above the roof was a platform from which a good view could be had of the surrounding region. The hut was thatched with grass.

In Buschan's *Illustrierte Völkerkunde* (Vol. II, 1924, p. 58) there is a picture of two houses built in one tree by the Kai people in the Bowutu Mountains south of the Huon Gulf. The lower one was elevated about 13 m and the higher one about 21 m in an approximately 23 m tall tree. Both houses seem to have been one-storeyed.

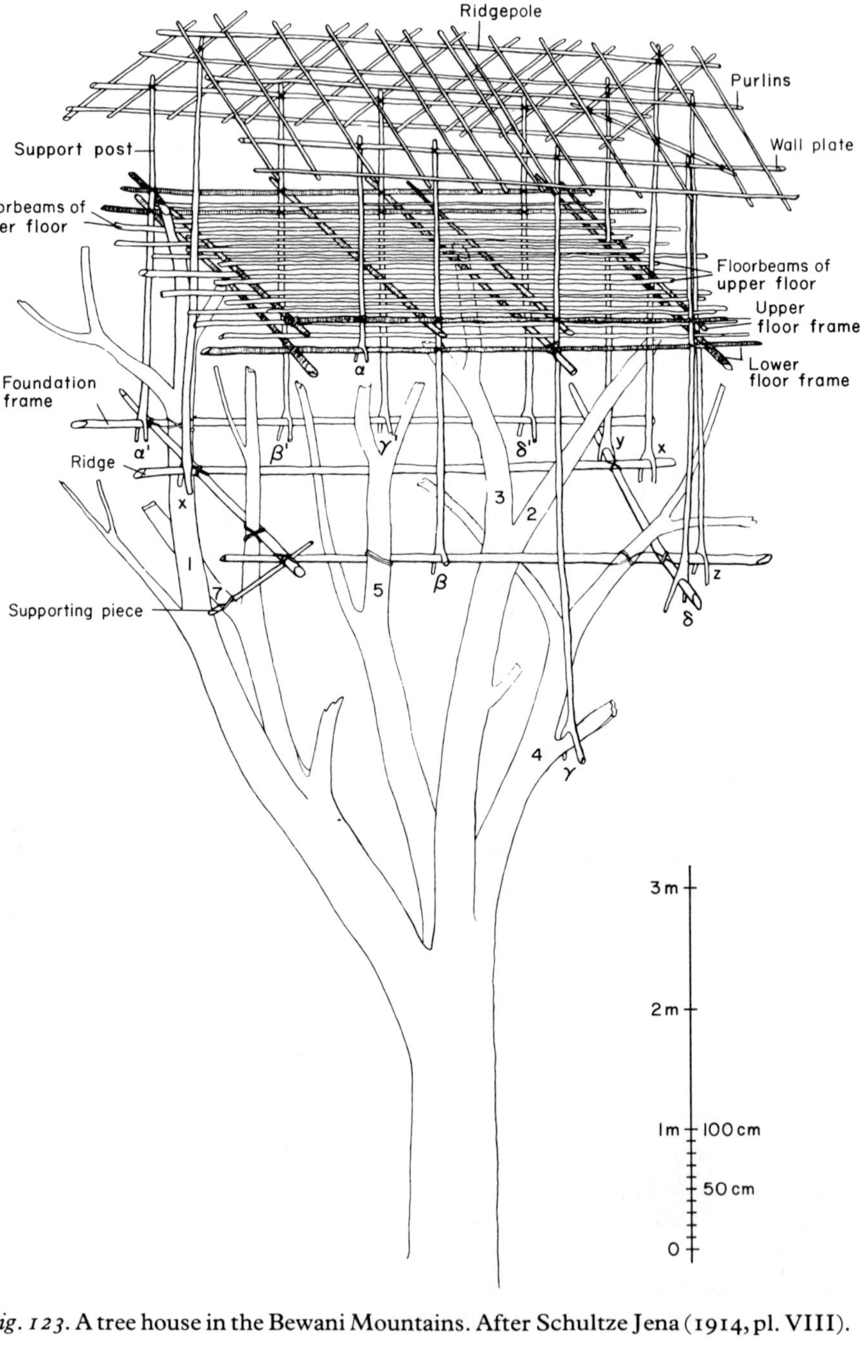

Fig. 123. A tree house in the Bewani Mountains. After Schultze Jena (1914, pl. VIII).

Fig. 124. Tree house in a Koiari village in the Astrolabe Mountains behind Port Moresby. Photograph by J. W. Lindt in *Picturesque New Guinea* (1887, p. 36, Pl.xiv).

Tree houses have also been described from South-East Asia (Buschan, Vol. II, 1924, pp. 781, 790, 807, 825, 830, 833), from south China (Buschan, Vol. II, 1924, p. 641), from India (Buschan, Vol. II, 1924, p. 541) and from South America (Buschan, Vol. I, 1922, p. 342).

Tent-shaped houses seem only recently to have entered the literature on New Guinea (Steensberg, 1973; Kauffmann, 1975, Fig. 106) and certainly none of them has been measured. When I visited Wewak in 1971, Michael Somare took me to Angoram and Ambunti on the Sepik river, as well as to Maprik and Aupik villages in the Prince Alexander Mountains. From the aircraft I saw several houses, the roofs of which stood on the ground, without vertical walls and lower behind than in front; from above they looked almost like pigs sitting on their haunches. Apparently they were all old and the style would soon disappear because new houses were being built in the modern style with substantial wooden walls and floors often raised above ground, resting on piles. I was able to photograph two of these old houses in Aupik but at the time I had no measuring equipment with me. I promised myself that I would return to the region during my next visit to Papua New Guinea but unfortunately I was unable to do so during my 1975 visit. I can only refer therefore to my observations of 1971.

As can be seen in Fig. 125, which shows the front porch and gable, these houses were constructed like a huge tent. Strong rafters slanted up from the ground towards each other and, where they crossed, were tied by cane lashings to the ridgepole about 4 m above the ground. Two purlins on each side, halfway between ridge and floor, supported the rafters on the inside but were not attached to them, being themselves supported on the inside by substantial timbers slanting upwards parallel to the rafters. As the drawing shows, the front gable sloped slightly backwards. It consisted of a screen of cane grass (or it could possibly have been young bamboo). A curtain of banana or palm leaves hung where the screen could be pushed aside for people to pass into the main room behind, which I was not allowed to do. The rafters were covered with thin stems of pitpit and above these was a thatch of leaves. The ridge was covered with palm leaves and some poles leant against them to stop them blowing off (Fig. 126).

In the front porch there was enough space for household work requiring daylight. Here, as seen in the sketch, a woman could sit working with her netbags and preparing food. A fireplace is shown

Fig. 125. Front porch of tent house, Aupik village, Maprik Sub-Province, Eastern Sepik Province. Drawn by the author, 1971.

Fig. 126. Tent houses in Aupik village, 1971: the house to the right shows a view of the rear gable; the other house shows a view of the front.

to the left and although this is near the roof, a protective layer of soot would soon build up above and the draught would carry away any sparks. The heap of stones shown at the rear would be used in the cooking pit. The whole scene looks attractive in its open simplicity. There was a similar, though lower porch at the rear gable as can be seen in Fig. 126. Although the picture suggests that the two houses preserved were fairly small, they were in fact 2·5 m high at the rear end, 4 m high in front and 8–10 m long.

An archaeologist might ponder the question of recognizing the site of such a house and detecting its construction. He would find no evidence for its roof in the surface of the ground; at best he might discover some ashes where the fireplace had been. Nobody would think that this fireplace had been near the front gable of the house and that it had been covered by a sloping roof quite near to it. Indeed, to an archaeologist the remains of the tent houses of New Guinea would be as unsubstantial as those of houses built in trees.

5. Houses with Wooden Walls in Irian Jaya and the Sepik region

Le Roux made excellent drawings of the wooden-walled houses constructed by the Dani people near the Baliem river, the Dém people near the Nògòlò-Rouffaer river, the Moni people in the Kemi valley and the Ekari people near the Paniai Lakes (Le Roux, 1950, pl. CXVI–CXVIII). The abundant illustrations in Gardner and Heider (1968) provide useful information about the timber buildings of the Dani villages visited by them between 1960 and 1963 while Heider (1970, pp. 252–268) analyses the construction of the houses.

According to Le Roux the Dani lived in round huts with walls of upright planks set in two rows, the inner ones covering the cracks between those in the outer row. In his illustration of a women's house (reproduced here as Fig. 127) there seems to be a wall-plate on top of the wall staves. After the wall had been built, four central posts were erected and from these posts rafters were inclined down to the wall. Above the central fire there was a baffleboard fixed to the four posts and carrying firewood for drying. Around the wall there were elevated platforms. The men's house was provided with a sleeping loft, and according to Heider's observations about 20 years later this sleeping loft could have its own fireplace. Also, according to Gardner and Heider, the domed roofs of the round houses were made of tall saplings stuck in the ground, bent over and lashed to the centre posts and to each other. This practice seems to have developed during the 20 years after Le Roux visited the area. Le Roux also described special long houses for cooking and for pigsties. Their walls were formed of horizontal planks slotted between uprights pushed into the ground as described previously for Yuku, while the roofs were constructed with central posts and a ridgepole roughly in the same way as at Yuku.

Like the Dani women's houses, the women's houses of the Dém, Moni and Ekari peoples drawn by Le Roux had no sleeping lofts. However, the floors were elevated 1 m or more above the ground surface, which is similar to the houses of the coastal regions, as well as the Swiss Lake Dwelling houses from the Neolithic. The elevated floor was probably no warmer than an earth floor because of the draught from below, but it was definitely drier which is important in the damp climate of the tropics.

The Moni men's house (Fig. 128) had a wooden floor just above ground level and a sleeping loft halfway up to the roof, each of them being equipped with a fireplace. On sloping ground the posts below the ground floor would be higher along one side than along the other, as in the Moni women's house shown in Fig. 129. Whereas the gables were always constructed with upright planks, the side walls were often made with horizontal boards, at least on the inside. In the Moni women's house firewood was slung against both the side walls and on the baffleboard above the fire in the interior room. There was also a fire in the front room and, again, firewood hanging to dry. In the family house (*Gezinswoning*) of the Ekari (Fig. 130) there were three fireplaces and firewood drying next to the walls and the rear gable.

The construction of houses in the Sepik region was similar to those described above though they were of much larger dimensions. Figure 131 (Schultze Jena, 1914, pl. x) illustrates such a house, 17 m long, 10 m wide and 8 m high from the ground to the ridge, in which about 50 people lived. The view shown is from the side facing the river, the main entrance being in the right gable. The gables were made of upright planks and the walls of horizontal boards. The roof was constructed around a central ridgepole and the house was stabilized by long, oblique braces running both transversely and longitudinally. The floor was elevated 4·5 m above the ground, and 11 hearths were distributed over it: two in the front hall near the gable to the right, two in the women's sleeping room on the river side of the central part, four in the men's larger sleeping room on

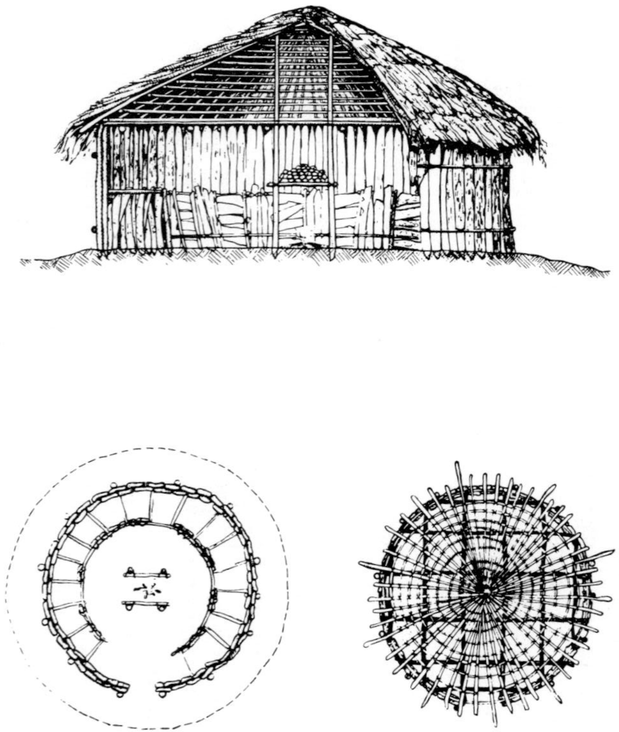

Fig. 127 (*left*) Women's house of the Dani people, Irian Jaya. After Le Roux (1950, pl. CXVI, I), with permission of E. J. Brill, Leiden. *Fig. 128* (*right*) Men's house of the Moni people, Irian Jaya. After Le Roux (1950, pl. CXVII, IV), with permission of E. J. Brill, Leiden.

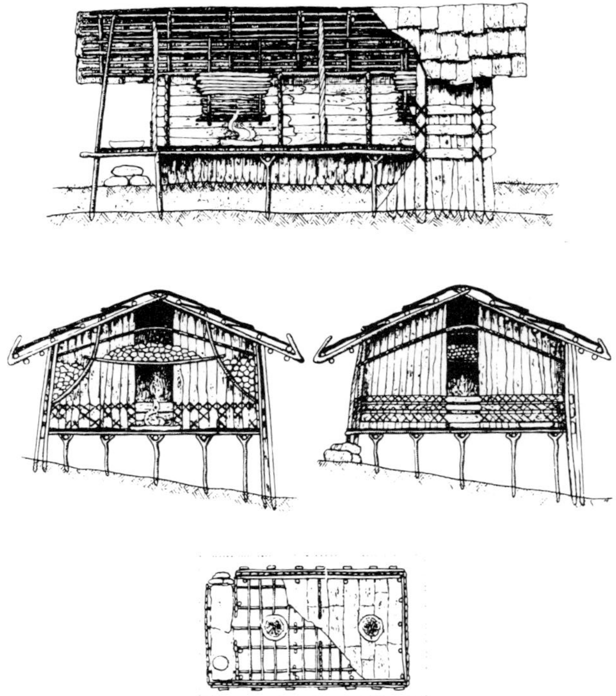

Fig. 129 (*left*) Women's house of the Moni people, Irian Jaya. After Le Roux (1950, pl. CXVIII, V), with permission of E. J. Brill, Leiden. *Fig. 130* (*right*) Family house of the Ekari people, Irian Jaya. After Le Roux (1950, pl. CXVIII, VI), with permission of E. J. Brill, Leiden.

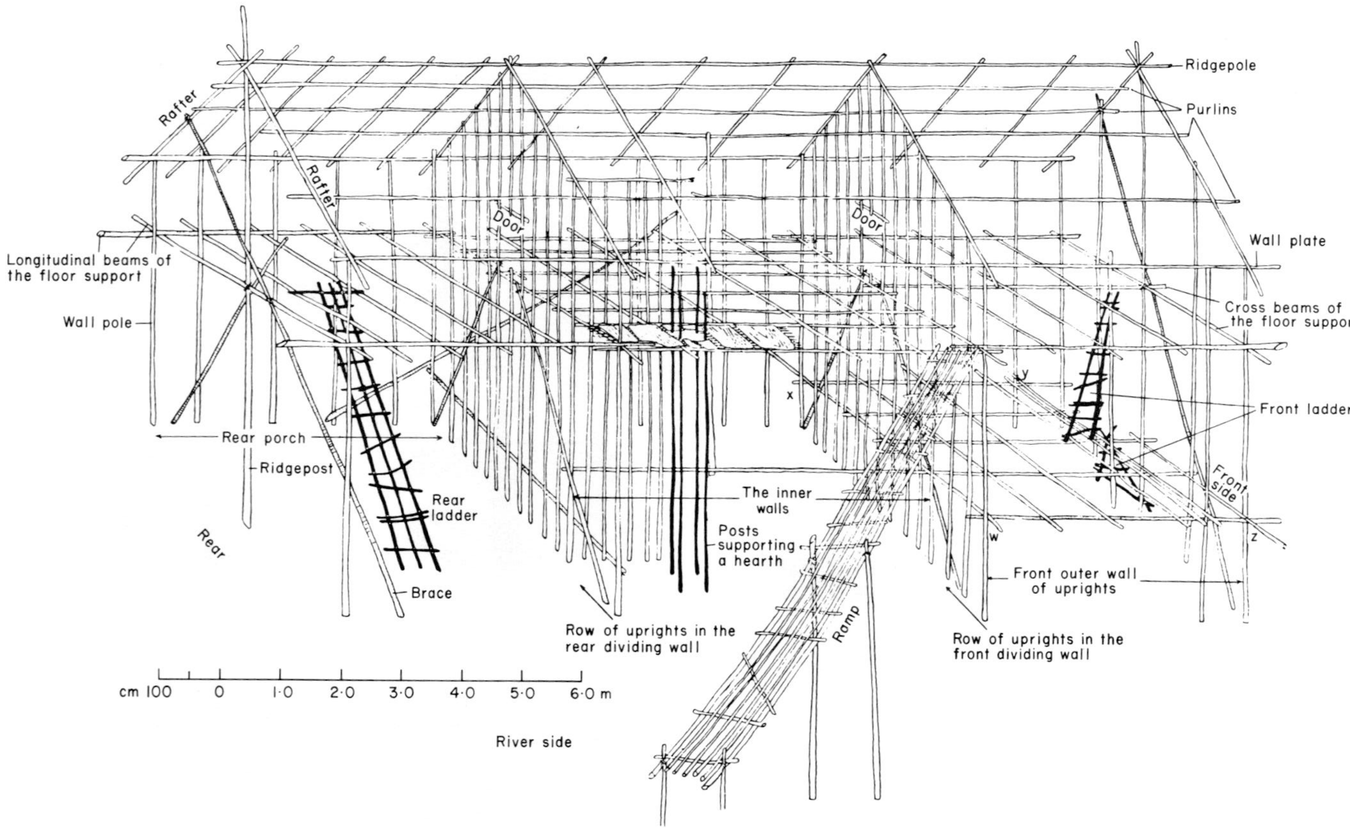

Fig. 131. Communal house in the Sepik region. After Schultze Jena (1914, pl. X).

the forest side, and three in the hall near the rear gable to the left. Baffleboards were placed above all the fireplaces. In the elongated cooking pits outside there were heaps of stones, 20 to 40 in each and all cracked from the fire showing that pig meat and occasional game had been prepared in the open air following the usual custom, while fish caught in the river could be roasted on the open fire. A 20 m long ramp connected the house with the river bank, crossing a zone overgrown with tall reeds.

6. Houses with Packed Walls in the Western Highlands

In 1968 I visited the Kindeng Forestry Station and Resettlement Blocks on the Wahgi river about 30 km east of Mount Hagen township. One of the settlers came from Wabag, about 80 km north-west of Mount Hagen. He was a well-to-do farmer and his household contained two women and seven children, and one of the latter spoke fairly good English. The women's house was built in the traditional Wabag style with an open porch in front and the roof top sloping upwards behind it. Inside the porch door there was a fireplace on which the woman heated stones for a cooking pit outside.

The most interesting aspect was, however, that the settler had started to build a new men's house behind the women's house. He was cutting the timber himself, but some helpers were engaged in the construction of a wall which was characterized by the packing of different kinds of cane grass between two rows of upright timbers. Building had started with the men hammering a row of uprights into the ground so that their upper ends were level. This formed the inside wall. Then some timbers were rammed into the soil in a row about 12 cm away to form the outer wall, but the space between each timber in this row was larger than in the inner row. After packing the gap between the rows with pitpit, strips of bark were slotted in between the packing and the outer uprights to form a kind of panelling so that rain water would not percolate through the wall (Fig. 132).

Figures 132–134 show the stages of binding the two rows of uprights together and the finished wall viewed from inside. Initially two stringers each made up of straight withes were tied one above the other to the top of the inner uprights with bast bindings. Lengths of strong rattan were pulled under the upper stringer, wound around both sets of uprights and tied firmly (Figs 132–133). Note in Fig. 132 the pressure being applied to the wall from the outside so as to keep the packing as tight as possible while the binding operation was under way. One man inside provided additional pressure in conjunction with the men outside as the "master binder" put his lashing through the packing (Fig. 133). When the binding was complete and the wall was firm and strong (Fig. 134), the upper ends of the outer timbers were cut down level with an axe. Then a central row of ridgeposts was erected to take the roof of the house.

The women's house had been lined in a different way outside, using matting instead of bark strips, and this fashion is now widespread. At Mendim, down a track off the road west of Kindeng plantation, the walls of the women's house were made in a simple way with uprights and horizontal and criss-crossing timbers packed with pitpit (Fig. 135). Both this house and the accompanying men's house were lined with plaited mats.

Sometimes grass alone was used to make a wall (Fig. 136). The upper part of the gable of this structure at the Kindeng Forestry Station was very much like ancient gable constructions of southern Denmark and northern Germany. In the top triangle there was a hole like the smoke-hole (*Eulenloch*) of the old Low Saxon houses without chimneys, and the gable was bent out as an eave to protect the lower part of the gable, just as on old houses of Lolland and Fehmarn. However, a direct German influence on building techniques here was unlikely, despite the German missions.

In some places one could observe a mixture of all the linings mentioned above—matting, bark and pitpit in one house (Fig. 137), which was a women's house situated in the Kum river swamps near Mount Hagen. The house was decaying and had been repaired above the door by pushing two pieces of modern corrugated iron into the thatch.

The making of pitpit matting was observed in the village of Alipe in the valley of the Kaugel river at the foot of the extinct volcano, Mount Giluwe. When Grith Lerche and I visited the village with Ron Lampert in 1968, as Dr Nancy Bowers' guests, she was going to have a house of her own built, and men from the village were occupied in making pitpit mats for its walls and floor. Some men peeled off the leaves and beat the stems flat with a wooden club on a stone or a stick (Fig. 138). The flattened stems were bent over into a ring and thrown to others who sat plaiting, producing a geometric pattern (Fig. 139). The finished mat was rolled up and carried to the house where it was to be used. The operation was conducted in a relaxed atmosphere; there was always time to smoke

Fig. 132. Applying pressure to the outside of the wall of new men's house as inside and outside uprights are lashed together with rattan bindings. Kindeng, 1968.

Fig. 133. Binding together the inner and outer rows of uprights of the wall of the new men's house at Kindeng.

Fig. 134. The packed wall of the new men's house at Kindeng from the inside, complete except for the levelling of the outside uprights.

Fig. 135. A women's house at Mendim, near Kindeng, 1968.

Fig. 136. A building with a thatched gable at the Kindeng Forestry Station, 1968.

Fig. 137. An abandoned women's house in the swamp near the Kum river, 1968.

Fig. 138. A man using a club to beat pitpit stems for matting. Alipe, 1968.

Fig. 139. A man plaiting the prepared pitpit shown in Fig. 138 into matting for house walls and flooring.

a cigarette made of home-grown tobacco and rolled in a scrap of old newspaper.

Another day Grith Lerche and I started the inspection and record of a family house in the village, which was convenient for us to do because the wife of Angu, the owner, was away with the children at a funeral in her clan. Most of the houses were spread over a large area down to the river, and Angu's house stood isolated from the others by stands of pitpit, cordyline, pandanus and other plants. Many houses in this region near the upper border of cultivation (just over 2000 m) are said to last for 10–15 years, and this one was at least ten years old. Grass was growing on the thatched roof (Fig. 141), and a bundle of bark from a large tree was wrapped up and stored under the eaves of the rear of the house together with a crude piece of trough, which Angu and his unmarried cousin Konumbu who lived with him used for carrying sweet potatoes from the garden below.

Because of the circumstances, we were even allowed to catch a glimpse of the owner's sleeping room in the north-western corner of the house, where he stored his personal effects. Konumbu lived in the other corner at the same end of the house, while Angu's wife and children shared the room in the opposite end. Figure 140 shows a plan of the house. In the large central room there was a fireplace containing ashes surrounded by protective stones with a baffleboard above. Between the fireplace and the front wall there was a cooking pit, and the roof above this and the fireplace was black with soot. Figure 142 shows the wall dividing the main room from the woman's room. The wall-plate was formed from bundles of withes on which the rafters rested. The grass thatch did not rest upon laths, as in Yuku and the houses in the forest regions, but as Fig. 142 demonstrates, it was laid upon a bedding of twigs, probably in order to save wood. Timber was available from the forest on Mount Giluwe, but this was some hours walk away and the only means of carriage was by the men's backs.

Along the rear wall of the central room were seven stalls for housing the pigs at night. Each stall was separated from its neighbours by double rows of uprights to stop the animals biting each other (Fig. 143).

The house walls consisted of broad horizontal planks fixed between two rows of upright timbers. Inside the house the planks were visible all the way round, but outside they had been lined with horizontal stems of cane grass (on the left in Fig. 144), which had partly broken off between the two doors. These were both formed out of loose horizontal boards slotted behind an upright on both sides and secured by an obliquely fixed plank, in an arrangement only known to the owner and his family. One of the two doors is shown in Fig. 144.

The eaves protruded some way out from the walls (see Fig. 145); note the layer of twigs visible between the rafters. Some of the rafters protruded more than others because they carried the weight of the eaves, and on these rafters were placed two horizontal laths. Some utensils and useful materials were hidden in gaps in the roof construction in the darkness under the eaves. However, the empty carton of photographic film shown in the middle of Fig. 145 was acquired from Grith Lerche.

A row of stones was laid along the foot of the walls, and although at the front of the house they were visible (Fig. 146), along the gables and rear they had been covered with mud to protect the house floor against water seeping in from outside. They probably also acted as a kind of stone drain, even when embedded in the mud. Rows or narrow zones of stones unearthed during archaeological excavations do not necessarily imply that wall posts had stood on them. For example, in Store Valby, Zealand, a pavement alongside one of the farmhouses was designed to prevent water from the hillside reaching the wall. This pavement diverted the water to the side of the house where it could flow downhill (Steensberg and Christensen, 1974, pp. 309, 311).

Not far from the house at Alipe there was a small secluded two-storied hut raised above the ground on substantial upright timbers with slender poles inserted between them (Fig. 147). Below was a shed for the pigs. Above were other poles between which a bark lining was interwoven. This was a special hut for use by females during menstruation and after childbirth. Under such circumstances women are not allowed to cook or handle food for the men for five days and they have to sleep apart from their families (Strathern, 1972, p. 166). That is the reason why this small hut was built in a solitary place, surrounded only by pitpit and scrub.

Apart from family houses, such as Angu's house described above, there are also men's houses in the traditional style in Alipe. Kupala, the man who demonstrated the use of the short paddle spade for mound-building, lived in a house with a porch in front of the entrance (Fig. 148) but with walls constructed in the same way as in Angu's house. When in 1933 or 1934 Kupala met one of the first white men

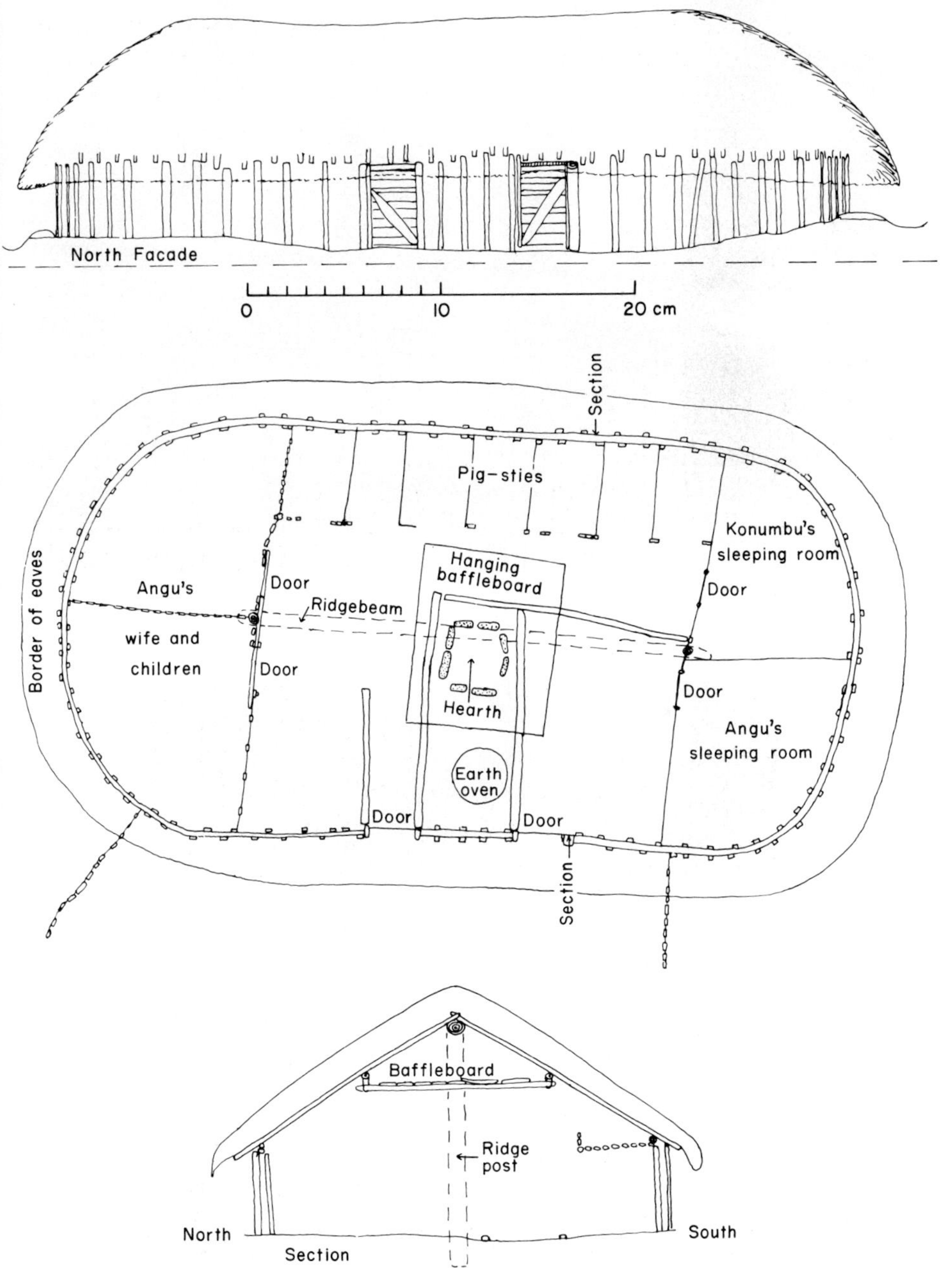

Fig. 140. Plan, front elevation and cross-section of Angu's family house, Alipe. Recorded by Grith Lerche and the author, 1968.

Fig. 141. The ten-year-old family house of Angu at Alipe in 1968, the slopes of Mount Giluwe in the distance.

Fig. 142. Constructional details of the roof, described in the text, are still visible in this part of Angu's house, despite the heavy deposit of soot from the hearth and earth oven.

Fig. 143. Pig stalls along the rear wall of the main room in Angu's house.

Fig. 144. One of the doorways at Angu's house blocked with horizontal planks, secured by a diagonal plank. To the right the pitpit lining on the outside of the wall has partly disappeared.

Fig. 145. View under the eaves above the locked door of Angu's house.

Fig. 146. Stones covered with mud along the foot of the walls of Angu's house protected the inside from flowing water. Those along the front of the house near the door had lost their mud covering.

Fig. 147. Hut for women during menstruation and after childbirth. Alipe, 1968.

Fig. 148. Kupala's house at Alipe, 1968.

Fig. 149. The guest house at Alipe, 1968.

to visit the Hagen region, Jim Taylor, he had not long been married and had small children, so in 1968 he must have been in his sixties. He is now a well-to-do man and one of the most important people in Alipe.

As guests in the village Grith Lerche, Ron Lampert and I were privileged to stay in the small hut shown in Fig. 149. This hut was constructed in the traditional manner with a post at each gable carrying the ridgepole. The walls were made of uprights with a bark lining retained by pitpit. A modern door had been brought from elsewhere to block the entrance as required. There was a fire-pit just inside the door, and although it was full of muddy water when we arrived, a boy emptied it and made a fire for our convenience. Every morning at daybreak he came with firewood and a burning stick and made a new fire, so that we could heat water for our morning toilet. But we had to lie stretched out on our long platforms until the smoke dispersed.

The houses of the Bomagai-Angoiang in the Ndwimba basin on the north side of the Bismarck Range are not constructed very differently from those just described (Clarke, 1971, pp. 100–106). Two or four large pointed posts are driven into the ground as a support for the ridgepole. Split prop roots of pandanus trees, sloping down from this beam to the walls as rafters, support laths of either bamboo or pitpit stems attached to them by rattan vines or something similar. Many plants are used as roofing material, for example leaves of bamboo, pandanus, banana and the like or bundles of pitpit. The walls are constructed of two rows of small posts driven into the ground, packed in between with long leaves of pandanus or strips of bark and lashed together with vines, like those described from Kindeng. The walls are seldom more than 1·2 m high, and the entrance rarely more than 0·9 m in either direction. The floor is often covered with peelings and spat-out fibres of the sugarcane that people eat as snacks and as a rainy-day food. But the men's houses have no porch in front as in the Hagen and Kopiago regions. Doors are often closed with planks, and the fireplace is a circle of rocks around a depression as described for other regions above. As in the family house of Alipe there may often be a cooking pit inside near the door for use in bad weather, but normally people use an outdoor pit for their cooking. The fire is kept burning all night because of the rapid cooling down after sundown at this altitude, and as usual firewood is stacked on a baffleboard above the hearth to dry.

The layout of the houses is different from those in the other areas described. One men's house (plan in Clarke, 1971, p. 105, Fig. 5b), in which two to three men and several boys slept, had a platform behind the fireplace in a slightly widened end of the house with a semi-circular gable. The normal women's house had a kind of porch in front and a platform in the semi-circular rear end. The women's houses as well as the family houses had separate compartments for the pigs from which a fenced passageway led outside the compound.

From their intimate knowledge of the Gende people living between the Bismarck Range and the Ramu river, Aufenanger and Höltker (1940, pp. 18–22) published one of the most detailed descriptions of house building available. Here, as with the Dani and other New Guinean peoples, the men's house was circular in plan, only 4·8 m wide and 2 m high in the centre. The entrance was flanked on both sides by five posts and when the horizontal planks which formed the door were not in position, a curtain of dry banana leaves prevented people looking in. The leaves were fastened to a piece of rattan above the doorway, so that people had to draw them aside to enter, rather like the chain or bead curtains in shop doorways in other hot countries.

Inside the entrance the fireplace was surrounded by stones as usual, but it was noticeably cramped for space. Behind the fireplace there was the large central post of the structure with two smaller ones flanking it on either side and three yet smaller ones beyond each of these; the whole forming a close-set row in the centre of the house at right angles to the entrance. Behind it was a wickerwork sleeping platform made from bamboo strips and raised 30 cm above the ground. Along the wall beyond a semi-circular pole, forked sticks served as a head rest when the men were sleeping.

In other parts of New Guinea men had portable head rests. In the store of the Kuk Tea Research Station we found a piece of wood, 45 cm long and oval in section, though flatter on one side than the other. It had been found in the swamp together with other wooden articles, for example the diagonal cross-bar of a plank door (cf. Fig. 144) and a club, but we could not guess what it might have been used for. However, when the old men from Mbukl saw it, they immediately recognized it as a wooden pillow. When during the hafting of the stone axe at Kuk Ul became drowsy, he simply stuck the pointed end of the axe handle in the ground, cradled his head between socket and blade and fell fast asleep (Fig. 150). Wirz (1952) says of the Enga that they did not use special head rests, only crude blocks of wood, perhaps of the same kind as the wooden pillow from Kuk. Inland from the north coast west of Vanimo on the other hand, Schultze Jena (1914, p. 34) found head rests furnished with V-shaped legs. Head rests with legs were also known in other parts of New Guinea, and in some areas they were decorated (Buschan, 1922–1925, Vol. II, p. 86). Head rests were also common in other parts of the world, in Africa, including dynastic Egypt, in east Asia and in Indonesia and Oceania.

7. Comparisons in Time and Space

As noted in the introduction to this chapter, a house reflects the material available in its environment, the kinds of activity performed by its inhabitants and their number and social status. The construction of the roof may reflect all these aspects but particularly the first. The influence of local materials is immediately noticeable in, for example, the different geographic regions of Britain, the slate roofs of Cornwall, Devon, Wales and the Lake District; the slabs

Fig. 150. Ul from Mbukl using his stone axe as a head rest after completing its hafting. Kuk Tea Research Station, 1975.

of oolitic limestones used in Northamptonshire, Oxfordshire and Gloucestershire; the roofing tiles of the Fen country and, moving northwards, other types of roofing material culminating in the turf coverings on the "black" houses of the Scottish Isles. The latter evolved as a consequence of the wet and windy climate, and such roofing was also common in Jutland in the Iron Age.

On small islands like Laesø in the Kattegat, seaweed was a traditional thatching material, which dried out into a compact sort of felt that might almost be termed "everlasting" (Engqvist, 1944). The lower rows were made of bundles of seaweed, the upper ends of which were each twisted into a kind of rope and simply looped over the laths. Higher up on the roof the seaweed was loose and the lower rows were made especially thick to stop the loose seaweed sliding down, just as in parts of New Guinea there was an extra thick bedding of thatch above the eaves. In both cases the thatch was without bindings or pegs.

Loose thatching was probably more common in former times than normally assumed. A striking feature in all areas is that roofs made of loose straw or grass are rather low, i.e. the pitch is below the normal angle of 50–60° for fixed straw and reed thatch. The house measured in Alipe had a pitch of only 30° and the roofs of the Yuku houses were pitched at about 20°. In the Hebrides the angle was 40°, in the Orkneys 30° and in the Shetlands 45–50°, according to Roussell (1934, cf. Fenton, 1978, Fig. 40, pp. 175–190). Though straw thatch is said not to be waterproof below 50°, I know of thatched roofs in northern Jutland with a pitch of 40°, because such low roofs are less exposed to the strong north-westerly winds of this region.

Loose roofing was used in south-east Europe as well, a reflection of the farming practice in the area. The corn was not threshed with flails, but the grain separated from the straw by threshing-sledges or horses' hooves (Schier, 1966, p. 80). Consequently the straw became so entangled that it just could not be tied to the laths. It was simply heaped on top of them and held down by branches and logs. Hay, ferns, rush and bark were used in the same way, similarly to the practices in New Guinea.

In old Europe tiled roofs became a symbol of social progress, but because many thatched houses were interspersed destructive fires repeatedly swept through crowded settlements. Sometimes a man's profession was reflected in the roof of his house: a baker could hardly thatch his bakery, and a blacksmith and a potter also needed a fireproof roof. In the New Guinea Highlands on the other hand, such professions never existed, and a tiled roof would be of no use for a house built only to be moved and rebuilt every three to ten years. However, roofs of corrugated iron have recently been introduced in the towns, and from them rain water is channelled into tanks of the same material, not only providing water for the household but also providing a visible sign of social status.

From prehistoric times walls of timber and clay were used everywhere in Europe. However, in the early civilizations of Greece and Rome they were gradually replaced by substantial walls built of stone blocks or bricks, at least in official buildings. This fashion spread into the Roman provinces, but the art of making bricks and tiles was not introduced into Denmark, for example, until about 1150 A.D. Walls built of horizontal timbers, jointed at the corners, were known in the Swiss and north Italian Lake Dwellings and became fashionable in Central Europe in the Hallstatt period. But in northern and eastern Europe, despite the huge areas of coniferous forests, dressed timber only began to replace the traditional mud walls during the tenth century. A.D. In rural Denmark half-timbered houses, mud-plastered walls and walls made of horizontal oak planks slotted into uprights at short intervals predominated until the eighteenth and nineteenth centuries, though in the towns brick-built houses had been common since the Renaissance period.

It is not, therefore, long since most peasants in Europe lived under similar conditions of housing to those of the gardeners of the New Guinea Highlands today. It is only to be expected, however, that the buildings which encompass the activities of the modern Papua New Guinean at home will be adjusted to the new and more specialized structure of his lifestyle, as has happened in Europe during the last 200 years. Consequently it is time anthropologists and archaeologists intensified their efforts in recording and interpreting the layout and construction of traditional houses and settlements. Otherwise we shall lose the real meaning of the excavated settlements on which the writing of the country's long history depends.

It is also timely for the Papua New Guinea authorities to consider the preservation of traditional architectural styles, not only magnificent and impressive communal buildings like the *haus tamboran* of the Sepik, but also the dwellings of the ordinary people such as those presented here from rural areas like Puya-kira'go and Yuku. In this way future generations may not only learn how their forefathers adapted with skill and intelligence to a greater variety of tropical environments and exploited their resources, but also perhaps be helped by the achievements of former generations to meet the challenge of development in an appropriately Papua New Guinean way.

5 | Fire and Cooking

1. The Technique of Fire Making

It is surprising, when one considers how easily a new fire is made, how much effort is spent in keeping a fire going and transporting it in the form of a burning stick over considerable distances. Perhaps the custom is dictated by tradition, or it may be easier physically to fetch a burning stick from the fireplace than to prepare a piece of wood and a strip of rattan or bamboo for a fire-saw, even when the burning stick has to be carried a long way. But if a hunter is roaming far from home, or if he is caught by a thunderstorm and cannot reach his house before night, he has to stay in a temporary hut or a rock shelter until the next dawn, in which case he may light a new fire to warm himself during the night and to roast some vegetables or small game for his evening meal.

Across the long arc from India through Indonesia to New Guinea and Australia drilling or rubbing to produce fire is uncommon. Here the fire-saw is the predominant tool of ignition. The technique is still remembered in many places of New Guinea, and I had Irari and Dundaki demonstrate it so that it could be photographed and filmed in detail.

From a branch 3 cm thick Irari cut a piece 32·5 cm long and pointed it at one end. At the other end he made a cleft about 20 cm long and opened it up with a wooden wedge about 1 cm thick, so that there was just space enough for the tinder. This consisted of some very dry and porous material and it was placed in the cleft behind the wedge (Fig. 151). Then Irari took a strip of rattan 40 cm long and split it into two halves so that each piece was flat on one side

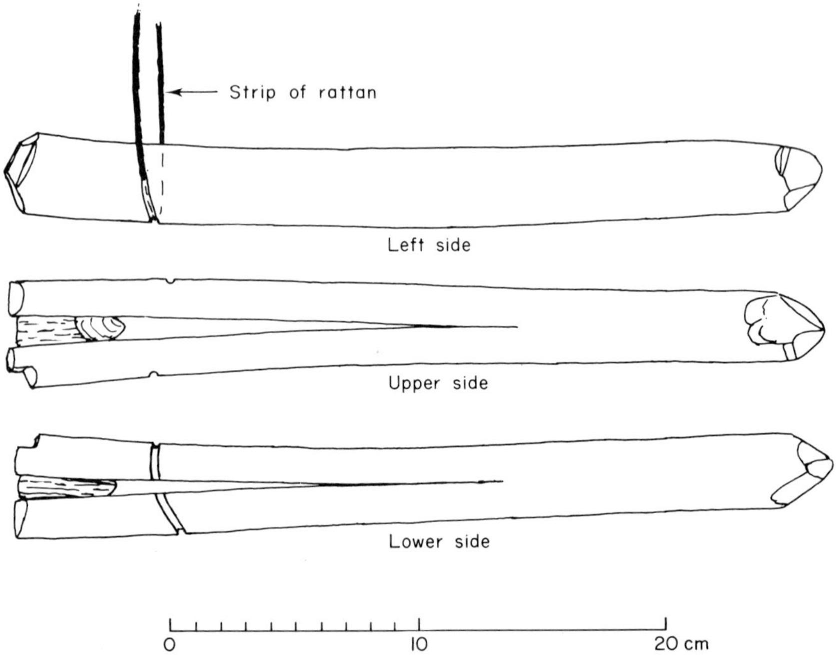

Fig. 151. Fire-saw used by Irari at Puya-kira'go, 1975. Drawn by the author.

and rounded on the other. He put one of these under the tinder stick at the point where the tinder was wedged, with the round side facing the stick and some dry leaves placed beneath. He placed his left foot on the pointed end of the tinder stick and his right foot on the other and grasped the two upturned ends of the rattan between his thumbs and forefingers. Then he began to saw with a short but strong up-and-down movement, his body bent forward and his legs straight. After about 15 seconds the string broke as a consequence of the heat produced and a dark-coloured groove due to singeing could be seen on the underside of the stick. The powder produced by the sawing ignited and settled in the top of the tinder.

Irari went down into a squatting position, blowing gently on the ignited powder and manipulating the stick with his left hand, until the tinder fell down on to the dry leaves below it. Still blowing on the tinder, he put the stick aside and lifted the leaves with his left hand until they were about 20 cm from his mouth. Slowly he moved them nearer to his lips until they were only about 5–6 cm away, still blowing gently. When the first smoke appeared he lowered the leaves to about 15 cm from his face so that he could still breathe. A handful of dry straw was then placed on top of the embers, soon to be followed by another. After this the smoke became so irritating that he had to keep turning his face away, but by maintaining the blowing he produced the first little flame. At this point the bundle was lowered to the ground and Irari started to feed it with small chips of dry wood. He rose from his squatting position, still bending the upper part of his body above the small fire and blowing constantly. A branch with dry leaves was immediately consumed by the fire together with a handful of dry grass. Now he could stop blowing. The flames flickered, fed by more chips and subsequently with wooden sticks. About 2 minutes after the string had broken there was a substantial fire.

In a skilled man's hands the art of making fire by sawing is easy and quick, but unskilled young men may try more than once and break several rattan strips before they succeed in producing fire. Irari was a man who succeeded in whatever kind of work he undertook, and, as Salisbury (1962, p. 57) puts it: "To be skilful is to be important," a proposition, however, that circumstances may qualify (e.g. White *et al.*, 1977, p. 390). Although at the time of my visit Irari was probably still in his thirties, it was inevitable that he would soon be recognized as an important man.

Such men should also be provident, calculating the effect of their actions on the future. In New Guinea I learnt that the need for good planning is everywhere. Men have to calculate where future swiddens should be located and at what time slashing and burning should be started so that one emptied plot can immediately be replaced by another. When their sons are born they must plant fruit trees and other useful trees which will be ready for use when the boys have grown into men. In fact the men around Irari also proved to be skilled planners of time in the short term too, so that I never wasted any time when I stayed in Puya-kira'go.

2. The Provision of Fuel

What surprised me when passing through the secondary forest was the large number of dead, ringbarked or pollarded trees (Fig. 152). Most of them were in recently abandoned gardens, though a significant number were seen in mature forest. In Chapter 1 I described how in clearing the bush for cultivation some tall trees were always left standing, often ringbarked so as not to compete with the garden

Fig. 152. Abandoned swidden with dead trees left from the previous clearance serving as a fuel reserve. Yuku, 1975.

crops for nutrients. They gradually die and ultimately provide a supply of dried-out firewood. The stocks of fuel so often mentioned in the description of the houses (Chapter 4) derive from such trees. Dr Takács of Budapest tells me (personal communication, 1976) that in former times in western Hungary the dead trees in the swidden were often left standing for 10–15 years or even longer and provided dry timber not only for firewood but also for buildings and tools.

Irari said there was always a great demand for dry firewood in the Puya-kira'go area because of the damp climate and cold nights. Based on the amount of split wood used every day in the fires of a men's and women's house over a period of 11 days in September 1975, I calculated that every family needed about $1 m^3$ of wood per week. This is more than double that required by Danish peasant families in some parts of Jutland in the 1920s for a whole farm. Here all of the fuel was dug out of bogs in the form of peat and the requirement of a whole farm for a year was about 20,000 turfs, each 1000 of which were the equivalent of about $1 m^3$ of beechwood.

However, by the beginning of this century all Danish farms had kitchen ranges and stoves of cast iron and all rooms were provided with tightly fitting ceilings, doors and windows that prevented heat escaping. Also the walls were much better insulated than those of Papua New Guinean houses. On the other hand, the climate is colder, in summer as well as in winter, and it is certain that before the introduction of plank ceilings into Danish farmhouses, the amount of firewood burned per year was much greater than the figures quoted.

3. The Pig Feast at Yaku

Irari and his friends celebrated the independence of Papua New Guinea, and my own visit to their territory, which coincided with it, with a feast in which they killed 14 pigs. As the most important men had to walk to the Patrol Post at Kopiago to attend the official independence celebrations on 16 September 1975, our local feast was held the day before. Activities began just after daybreak and when I arrived at 7:30 a.m. most of the pigs had already been killed.

Because the participating people were now Christian, the spiritual side of the ceremony had changed from what it had been in the pre-conversion era, but the practical part of the feast was traditional, and that is what will be described here.

The method of killing a pig was for its owner to hold it in his arms while another man knocked it out with a sharp blow on the forehead with a wooden club, rather like a contestant in a boxing ring (Fig. 153). The pig was then dropped on the ground and beaten to death. The preliminary blow ensured that the pig felt little pain. When an animal had been killed, it was singed in a fire close to the spot where the big cooking oven would be dug, emerging brown and smooth and free of bristles. A great deal of firewood was being split and chopped into convenient sizes, shorter ones for the singeing process, longer ones for the bottom layer of the cooking pit.

A place for preparing the carcasses had been selected alongside the bush. Here the "master butchers" prepared an "operating table" for their pigs, covering the ground first with a layer of ferns and then a layer of banana leaves, on top of which the singed bodies of the pigs were placed. A bamboo knife (Fig. 154) was sharpened by detaching a sliver along one edge with the teeth and biting it off, the new edge being as sharp as a razor blade. One of the men, Hayapa, occasionally used a steel knife, but Palapa, Hemoko and

Fig. 153. Before the pigs for the Independence Day feast at Yaku were slaughtered, they were knocked out by a blow on the forehead from a wooden club.

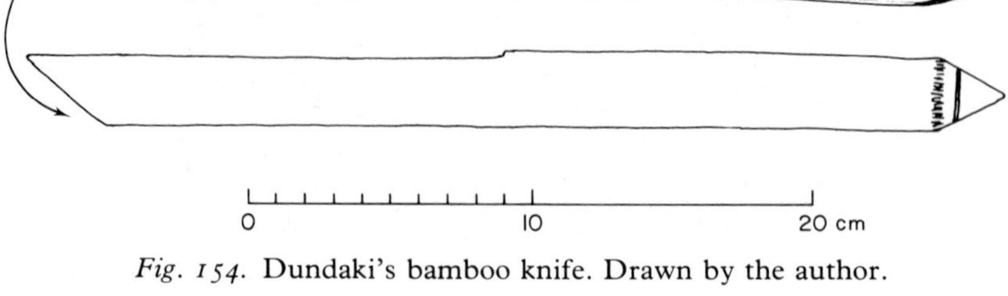

Fig. 154. Dundaki's bamboo knife. Drawn by the author.

Dundaki, as well as Irari, all used bamboo knives exclusively. The knife shown in Fig. 154 belonged to Dundaki. Incidentally, these knives are also used for trimming hair and beards.

Figure 155 shows how Dundaki opened up the belly of a pig by making a cut on each side. He worked cleanly like a surgeon, though dressed in his ceremonial garments with the waist-strings around his hips that are characteristic of this region, as opposed to the bark belts found further east in the Highlands. In front he wore some decorative fringes. His leather belt was a modern acquisition, but his bone dagger was a traditional tool. Around his neck were strings of coloured beads with a brilliant plate of pearl shell in front. His

Fig. 155. Dundaki opening up the belly of a pig with his bamboo knife.

head dress was the most valuable and magnificent part of his outfit: a cap made of hair bordered by possum fur and decorated with feathers and tails of birds, particularly a long beautiful tail of a bird of paradise. The pearl shell and the head dress demonstrated his wealth and importance.

While the intestines were being removed from the pigs, the men deftly wiped up the trickling blood with fresh fern leaves. Then the women took the entrails away, carrying them in their net bags to a small creek nearby (Fig. 156). Here the intestines were emptied

Fig. 156. The women carried the intestines in their net bags to a nearby creek for cleaning.

Fig. 157. Yopa teasing three of the dogs with a small strip of pork. Note the boar's tusk in his nose and the traditional bone dagger tucked into his waist-strings.

Fig. 158. Some smaller sides of pork were hung temporarily over a rail while the men in the background were occupied in singeing other slaughtered pigs.

and washed; no part of the pig was wasted, everything was cooked and consumed.

In the meantime Yopa teased the dogs by letting them smell a small strip of pork. Though these well-trained animals had drooling mouths, they dared not snatch it and in the end the prize went to the smallest dog (Fig. 157). Yopa was not as elaborately dressed as Dundaki, having a necklace of glass beads wound around his head and another of seeds hung around his neck. An arm-ring of woven fibres adorned his left arm and the traditional bone dagger was tucked into his waist-strings. The boar's tusk in his nose was part of his everyday decoration.

During the butchering the larger portions of meat were placed temporarily on fern leaves close by and some sides and legs were hung over a horizontal pole supported at the ends by two forked sticks (Fig. 158).

The cooking pit for the 28 sides was dug at the traditional spot, where the soil was mixed with charcoal from previous ceremonial feasts. One of the old men broke up the earth with a digging stick about 1·5 m long, the lower end of which was bevelled on one side. He used the stick with the bevel away from him, digging in the usual way with his front hand in overgrip and his back hand in undergrip. The loose earth from this digging was removed by another man using a steel shovel.

The oven pit was about 25–30 cm deep, about 8–10 m long and 1·25 m wide. It was lined with long pieces of firewood, and when nearly all the firewood had burned away, the men took the remaining pieces and made a new fire on the edge to heat stones to use for cooking. When these were scorching hot, some of them were transferred with wooden tongs (Fig. 159) or bare hands into the pit, where they were covered with banana leaves and ferns to form a bedding for the sides of pork, which were arranged in two interlocking rows. Another layer of banana leaves and ferns mixed with herbs to give flavour to the pork was added, and more hot stones were thrown on top. A further layer of leaves, this time of the *andukukone* plant, apparently identical with *Schefflera machrostachya* (according to Dosedla, personal communication) was added (Fig. 160) and subsequently the pit was covered with grass pulled from the surrounding area. An atmosphere of joy and excitement was maintained throughout this activity but it all had to be done in a hurry, because shortly after 1:00 p.m. the rain had begun to pour down, while thunder rumbled against the hills.

Fig. 159. The hot stones were transferred from the fire to the earth oven at Yaku with wooden tongs. Drawn by the author.

The pit was finally covered with earth so that no steam could escape (Fig. 161). The ground became wet and muddy and some of the men went into the nearby men's house and squatted by the fireside, chatting and smoking. The women sought shelter in the Catholic chapel on the other side of the cooking place.

Several other activities had been going on at the same time as the construction of the main oven. While the stones were being heated in the fire, the men spiked small bits of pork on pointed sticks, roasting them for a moment and eating them from the stick, rather like a Middle-Eastern kebab. Small pieces of pork cut off from the whole sides were wrapped in leaves and cooked with sweet potatoes and taro in a special earth oven to the right side of the men's house. Other pieces of pork were cooked wrapped in the cleaned intestines. Another oven was heated just to the left of the men's house and here my contribution to the feast, some rice and tinned fish, was cooked. The women had used this oven the day before to cook sweet potatoes so that a store of basic food would be at hand.

After the thunderstorm was over, the men went to a hillside behind the chapel and amused themselves and the children with an archery competition. A short pole was put up on the other side of the small valley about 100–150 m away. Few of the arrows went this far and only one of the young men hit the target. The bows, called *kendei*

Fig. 160 (*left*) The hot stones on top of the pork were covered by leaves of the andukukone plant. *Fig. 161* (*right*) Steaming in the rain of a thunderstorm, the oven was finally sealed with earth.

in the local language, are excellently made, the best of them from wood of the black palm; the strings are strips of bamboo scraped with a chert flake. The men used the deadly *mali* arrow furnished with a long sharply pointed bamboo head, but they also had some barbed arrows called *langa*, fashioned with large barbs along one side, like the bone harpoons of the Ertebølle time in Denmark. This arrow is used for shooting cassowaries and pigs, but not in war. The arrow called *kopi* was used in war as well as in hunting; it generally has a groove down one side, in which small cowrie shells are sometimes set. For shooting birds they use the *duku* arrow which has three prongs.

After about three hours of cooking the contents of the big earth oven were ready for consumption. The earth was thrown aside so that the steaming leaves were visible. These were removed with long digging sticks because they were too hot to be handled. Figure 162 shows the uncovering of the cooked pork which was allowed to cool until it could be held in the fingers. A portion was distributed to each family and some of this was taken aside by the women and children to be eaten separately from the men. I took my share of the pork to Puya-kira'go, where my interpreter and I had it for supper.

4. The Good Mother Feeds her Family and Pigs

The large cooking oven at Yaku was operated by men, but everyday food preparation and cooking is one of the women's duties, the others being to plant, weed and harvest the gardens and to look after the pigs and the children. Strathern's book (1972) on the role of women in Mount Hagen country is a commentary on the ways in which a man and his wife regard themselves as being in an economically reciprocal position. The men clear the land and chop firewood, bring in game from the bush, rebuild the houses, protect the settlement against enemies and organize the political life. But the heaviest burden of labour, and the most monotonous work, is placed upon the women.

Men do not think that women are as skilled and ingenious as themselves. However, much of women's work requires considerable dexterity, for example, spinning bark fibres into threads and making net bags. They also plan their work to avoid unnecessary effort; Clarke (1971, p. 165) describes how the women of the group with whom he was living would leave the bamboo tubes for carrying water at

Fig. 162. After about three hours of cooking the pork was ready to eat.

the water source as they passed on their way to the gardens in the morning, filling them on their return in the afternoon and carrying them home.

The division of labour between the sexes is a way of organizing necessary duties in a practical way. Parallel examples of economic reciprocity can be drawn from European peasant life, for example the symbiotic relationship between the manor owner of former days and his tenants. Modern historians are inclined to think of the latter as slaves of their master, but this is not really the case. In fact many Danish farmers were reluctant to take over their farms and total responsibility for the welfare of their families when the bonds of service were broken at the end of the eighteenth century. They knew that their masters provided security if crops happened to fail or if the livestock died. When they became their own masters, they had to pay their dues promptly to the state otherwise their farms were put up for public sale. The manorial system in part acted as an institution of social welfare in which all members were mutually responsible, though the incomes differed according to the input of capital.

Although the women carry the greater burden of the work in Papua New Guinean rural life, they know that they are indispensable. The household would suffer considerably from their absence, even though the men may be capable of tilling a garden alone, as we have seen in Alipe. The women may also, if dissatisfied, return to their own kin, though in this case the dowry must be paid back.

Women have always been nearer to the roots of life than men, bearing the children, nursing them and preparing their food when they are small, taking care of other members of the household as well as of guests. This was exemplified the last day I stayed in Puyakira'go when I was given a farewell party. Pokome came and presented me with a net bag she had made herself and a fowl, and Irari brought two more fowl. My cook and another young man cut the firewood while Pokome plucked and prepared the fowl. A fire was made above the oven pit and stones were piled up on it (Fig. 163). When this had nearly burned down, some of the hot stones were put aside (Fig. 164) and placed on another fire to keep them hot while the cooking pit was lined with banana leaves and ferns. Some onions were placed in the pit, and the man shown squatting on the right in Fig. 165 put the fowl on top of them. These were covered with more banana leaves and the whole covered with fern leaves. The hot stones were thrown on to the heap and distributed in an even layer, on which was put a layer of taro tubers and a small packet of leaves containing a fungus for flavouring. All this was covered with another layer of banana leaves and ferns. By now it had begun raining heavily, and though the earth oven itself was protected with a flat roof, supported on four posts, Pokome continued to stand outside in the rain and mud, scooping earth and organic rubbish into a heap to seal the oven in and patting it down firmly so that no steam could escape from the interior of the pit.

Pokome worked quickly and skilfully without taking the slightest notice of the pouring rain or of her watchers. She had her own

Fig. 163. On the eve of my departure from Puya-kira'go a meal was prepared in the local earth oven. The cooking stones were heated in advance on a fire on top of the oven.

Fig. 164. When the fuel had burned down some of the hot stones were put aside to be kept warm in a secondary fire for use in the upper layers of the earth oven.

Fig. 165. The man squatting on the right places three fowl in the earth oven. These were covered with banana leaves and ferns and an upper layer of hot stones was placed upon them. Subsequently the whole was sealed with earth.

dignity, a proud personification of her ancestors, an Eve or Mother Earth. Two hours after the oven had been covered Pokome served us a delicious meal.

5. Earth-Oven Cooking in New Guinea and Europe

In Alipe the men demonstrated the use of an earth oven for us. Their method was similar to that of Pokome at Puya-kira'go (Lerche, 1969). In this case a pit was made, about 50 cm in diameter and 20 cm deep, lined with leaves of cordyline and wild fig. The sweet potatoes and taro tubers to be cooked were put in the pit and on these was placed a layer of hot stones. These had been heated on a fire at the edge of the pit which had been ignited by sawing with a bamboo strip, similar to the practice at Yaku and Puya-kira'go. The hot stones were covered with leaves of cordyline and some aromatic plant, pitpit was piled up above these and finally grass turfs were laid on top and pressed down to prevent the steam inside from escaping. As only taro tubers and sweet potatoes were cooked, it took only about one hour to prepare the meal.

Clarke (1971, Appendix B) lists other aromatic leaves used in earth ovens in the Ndwimba basin, such as a tree fern, *Cyathea* sp., another fern, *Diplazium* sp., the leaves of which are cooked and eaten, *Pouzolzia hirta*, a garden weed, and *Solanum nigrum* which were rather harsh-tasting leaves. As well as earth ovens, the Bomagai-Angoiang have above-ground ovens some of which are used for the ritual cooking of the heads of male feral pigs, eaten only by men. They consist of a ring of stakes of *Alphitonia incana*, a plant of the Rhamnaceae family also used for firewood; these are bound together with vines and lined inside with banana leaves. Another above-ground but non-ritual oven is made by folding some flexible bark into a cylinder shape and lining it with banana leaves. The meals are prepared very much as in Puya-kira'go and Alipe, the ovens being lined with fresh leaves and the food put in alternately with hot rocks from the fire. After about an hour and a half the meal is ready, and a sauce is prepared from pandanus seeds, when in season (Clarke, 1971, pp. 109, 117). In the Hagen area plants like *Nasturtium officinale* (watercress), *Solanum brachypetalum*, *Commelina* sp., and *Justitia angustata* are gathered before the pigs are slaughtered and put into the earth oven together with leaves of fern and wild fig (Dosedla, 1974). In some areas an earth oven is often dug inside the house near the fireplace so that the stones can be heated on the hearth before being put into the oven. In all areas roasted tubers are eaten any time people feel hungry, and they are carried around in net bags by both men and women and taken as snacks alternately with pieces of sugarcane.

If pork is not eaten immediately it can be converted into smoked ham by hanging over the fireplace (Wirz, 1924, p. 97). The Bomagai-Angoiang people of the northern slope of the Bismarck Range temporarily preserve meat in this way, but the meat lasts only a few days because it is hung in bulky chunks instead of being cut into thin strips (Clarke, 1971, p. 169). The fat, which provides an important addition to the starchy diet, is also used for greasing the body, while in the past pig bone was used as a raw material for artifacts.

Parallels to the earth ovens of the New Guinea Highlands have been found in both Iron Age and Medieval settlements in Denmark (Lerche, 1969). A quasi-rectangular pit with rounded corners and a depth of about 20 cm was found during excavations at the medieval village site of Store Valby. The north-east corner of this was partly missing because a well had cut it away. The pit was packed with stones brittle from heat, and charcoal and fragments of firewood 6–7 cm in diameter placed along the north-west–south-east axis of the pit. The stones and charcoal were mixed in a layer above the pieces of wood, which were placed in two layers at the bottom. At the rear of the pit (Fig. 166), some marks in the subsoil could be seen, made by the round pieces of firewood which apparently had been

Fig. 166. An earth oven, probably of medieval origin, excavated at the village site of Store Valby, Zealand. Impressions of the round pieces of wood that formed the bottom layer of fuel are to be seen to the left and rear of the pit.

pressed down into the rather narrow pit in order to provide a bedding of fuel. After this had been ignited, it would have been covered by stones, as in the earth oven at Yaku. In the centre of the Store Valby pit some charred fragments of bone were found, arranged in the same north-west–south-east direction as the pieces of wood. This earth oven has not been exactly dated, but most pits in the surrounding subsoil were of medieval origin. The well referred to above belonged to farm buildings that had existed since the fifteenth century, so the pit must have been older (Steensberg and Christensen, 1974, Vol. III, pl. XIII, pit AP). Only 4 m to the north-east there was a pit containing pottery and flint tools from the last phase of the Middle Neolithic.

At a site of another medieval village, Borup in central Zealand, I excavated a large heap of brittle stones, 6 m in diameter, which apparently had been cracked by fire. At the south-eastern corner there was a thick layer of charcoal which was carbon-dated to the fifteenth century A.D. Other heaps of brittle stones were scattered over the site, but they have not yet been dated.

Lerche (1969) refers to a number of earth ovens from the Iron Age and illustrates one from south Jutland. Round wooden sticks had been aligned in one direction in the bottom of the pit, which was about 20 cm deep, and a layer of stones had been placed on top. She also shows profiles through ten earth ovens excavated at the Iron Age settlement called Nissehøj in south Zealand containing brittle stones and charcoal. From the old Norse-Icelandic literature we learn that cooking in pits was practised alongside cooking in vessels containing water heated by hot stones.

I am convinced that earth oven cooking was in fact a common practice in the countryside in prehistoric and indeed medieval Europe. Recently in renewed excavations at Borup I found an earth oven actually inside a house of about 1100 A.D., just like the New Guinea fashion. Wirz (1924, p. 91) observed that the ovens inside the women's houses of central Irian Jaya were lined with grass and leaves and the hot stones used for cooking were sprinkled with water from a bamboo tube in order to produce more steam. The heaped contents were covered with leaves, the ends of which were bent back over them and sealed with loose stones. It may well be that the same technique was practised in the Borup house, because the stones inside the oven had been shattered in a way consistent with the application of water.

It is unlikely that such a technique of cooking would be noted in written records, so it has to be recognized archaeologically. The oven is such a simple pit, however, that it can easily be overlooked when the excavator has no model to refer to.

Perhaps the meat cooked in such ovens was mainly pork, mutton and fowl. Cooking pits would also be useful when many mouths had to be fed at once, for example at festivals, as was the case at Yaku. However, no earth oven of comparable size to the large New Guinea ones have been described from Europe. The reason could be that at really large feasts people would serve a whole barbecued ox. Barbecuing and roasting were in fact common practices in town houses and castles in medieval Europe.

6 Conclusions

In the previous chapters hitherto largely ignored aspects of the techniques of husbandry and house construction have been analysed. Other aspects have deliberately been omitted or only touched upon, because they have been extensively treated elsewhere. Readers who seek an ecological context are recommended to treatises on limited geographical regions such as Pospisil's work on the Kapaukus of Irian Jaya, Wadell's on the Engas, and Vicedom and Tischner's on the Mbowambs of the Hagen area, or Clarke's on the Bomagai-Angoiangs on the northern slopes of the Bismarck Range. They may also find additional facts in the occasional papers or collective publications in the References on pp. 210–218, which of course contains only references that have been quoted in this book.

I have tried to concentrate on subjects that could inspire further research to throw light on the use of artifacts in ancient Europe. Experimenting with ancient artifacts has proved its scientific value in the Historical and Archaeological Experiment Centre at Lejre and in similar research stations elsewhere. The educational value of such experiments has also been recognized in the industrialized parts of the world. Theoretical teaching succeeds better if it is based upon practical insight, and an important part of practical insight depends upon knowledge of the functions of tools and implements and how they are adapted to their natural environment.

In this respect the local environments of New Guinea would offer ideal conditions for teaching students from the developed countries —not to mention the benefits of such mutual contact. However, an influx of such visitors would be disastrous for the local environment and culture, so we must find other means of transporting practical knowledge from New Guinea, as well as from other so-called "underdeveloped countries", to the students of the western world. One way to do this is to describe the material culture and its function in detail as I have tried to do in this book, so that teachers may transmit such experience to their pupils, imitating the use of tools and other equipment in experimental centres in order to learn the secrets of the material culture which was always the basis of spiritual life.

Tools and weapons in New Guinea look simple, but their use is very sophisticated. Their simplicity in design raises the question of whether they contain some of those "unimportant details" from which supporters of the "Vienna-School" of ethnology might deduce the existence of former cultural-complexes.

Are tools like the double-spade—which has parallels in the South African "spadoon" and double-spades of Jutland moors—relics of an ancient general network of culture, or are they simply examples of coincidence as independent inventions? Or could the complex and elastic way of hafting the stone axes in New Guinea and the Swiss Lake Dwellings be a result of an early general diffusion—not to mention the similarity of thin-bladed axes? Other similarities mentioned above are the ridgebeam construction in house building and the slate blade with semilunular incisions on both sides of a hafting-tang which is a parallel to the Syrian rope-traction ards of limestone or basalt. One might add that the smoking of tobacco or other kinds of leaves has spread to the remotest parts of New Guinea—could that be an independent invention in the island of New Guinea?

The question of whether or not a small detail is important arises time and again. Among a variety of arrows presented to me in Puya-kira'go was one with a number of barbs cut out of one edge while the other edge was smooth. It was called *Langa* in the local language. This type of arrow with very coarse barbs at one edge is not common in New Guinea. At least neither Pospisil nor Clarke mention it, but Heider has described similar types from the Dugum Danis in Irian Jaya. He deduces from Wirz's drawings in 1920 that there has been little change in arrow variations over time or distance, though there are considerable variations in other aspects of culture and language within the Dani area. One explanation is the constant warfare which ensures that arrows will be in constant movement from one region to another (Heider, 1970).

However, one detail in the *Langa* of Puya-kira'go has a world-wide significance: the hindmost barb is turned in the opposite direction to all the others, pointing towards the body it is to penetrate. Troels-Smith drew my attention to this odd position which has parallels not only in the Mesolithic cultures of Europe but also in the culture of the Eskimos of Greenland. Why did people so dispersed in time and distance cut the hindmost barb of an arrow in this odd position? Perhaps it was done so as to catch some of the muscles as in a trap and tear them to pieces thus producing an uneven wound more inclined to infection. Or it could prevent the arrow from penetrating too far or even going straight through the body.

In my opinion such traits are examples of independent invention, and they had a practical reason from the outset.

Nobody will deny the importance of the dissemination of ideas in culture, especially obvious in the history of arts. Elements of decoration are often the "unimportant details" that indicate a former cultural contact. But there is always a danger of being mistaken. Michael Somare once told me that most decorative elements of the famous art of the Sepik area express old clan-traditions and symbols which the boys learnt when they listened to the tales and legends told by the old men carving and painting those items. But might not two almost identical ornaments represent different traditions over a large area, or different traditions of culture if they were made in different parts of the world? The possibilities are many.

The structure of man's brain does not seem to have altered over the last hundred thousand years. This is the reason why people from different parts of the world and different cultures can communicate with each other without knowing much of the other's traditions and language—especially in material culture.

Despite this structural similarity within the human species, the inherited part of man's intelligence is different from being to being. Had Irari Hipuya been born in a western society he would probably have risen to a high position because of his inherited ability, his quick mind and excellent memory. In my home country people are just as differently gifted as those in New Guinea, though acquired knowledge often hides this fact. Some of us inherit mathematical ability, others are born musicians, and still others inherit a retentive memory which is not necessarily combined with great creativity.

Our creativity can be active in different directions, and though our brains generally speaking have been structured in the same way since prehistoric times, there will be differences partly because of genetic inheritance and environment. So the statistical possibility of two or more people getting the same idea seems to be obvious—

they could even express the same oral sounds or create the same design.

The brain is not a mechanical reproducer but a creative organ. The material artifacts reflect this genius of mankind and that is the reason why the study of material culture is so fascinating.

References

Alberti, L. (1968). *Ludwig Alberti's Account of the Tribal Life and Customs of the Xhosa in 1807*. Cape Town.

Allen, J. (1970). Prehistoric agricultural systems in the Wahgi valley—a further note. *Mankind* **7** (3), 177–183.

Allen, J. (1972). The first decade in New Guinea archaeology. *Antiquity* **46,** 180–190.

Allen, J., Golson, J. and Jones, R. (eds) (1977). *Sunda and Sahul: Prehistoric Studies in Southeast Asia, Melanesia and Australia.* Academic Press, London and New York.

Arrianos, Flavius. *Indice.* Cf. Arrian, English translation *Indice*, II (Book 8), (E. Iliff Robson, ed.). The Loeb Classical Library. William Heinemann, London.

Aufenanger, H. and Höltker, G. (1940). *Die Gende in Zentralneuguinea: Vom Leben und Denken eines Papua-Stammes in Bismarck-Gebirge.* Missionsdruckerei St. Gabriel, Wien-Mödling.

Avitzur, Shmuel (1965). *The Native Ard of Eretz-Israel: Its History and Development.* Man and his Work series of the Avashalom Institute for Homeland Studies, No. 5, Tel-Aviv.

Becker, C. J. (1945). New finds of hafted Neolithic celts. *Acta Archaeologica* **XVI**, 155–175.

Becker, C. J. (1949). Hafted Neolithic celts II: With observations on a new funnel-beaker type from Zealand. *Acta Archaeologica* **XX**, 231–248.

Becker, C. J. (1957). Den tyknakkede flintøkse [Engl. summary: The Thick-Butted Flint Celt: Studies on the Heavy Flint Celts of the Danish TRB Culture]. *Årbøger for nordisk Oldkyndighed*, pp. 1–37.

Bellwood, Peter (1975). The prehistory of Oceania. *Current Anthropology* **16**, 9–28.

Blackwood, Beatrice (1950). *The Technology of a Modern Stone Age People in New Guinea.* Pitt Rivers Museum, Occasional Paper on Technology No. 3. Oxford University Press, Oxford.

Blinkenberg, Christian (1898). Skæftede Stenalders Redskaber. *Aarbøger for nordisk Oldkyndighed og Historie* pp. 125–156.

Bowers, Nancy (1965). The ascending grasslands: agricultural practices and successional vegetation in the upper Kaugel valley, Australian New Guinea. Unpublished Ph.D. thesis, Columbia University, New York.

Bowers, Nancy (1971). Demographic problems in montane New Guinea. In *Culture and Population* (Steven Polgar, ed.), pp. 11–31. Schenkman, for Carolina Population Centre.

Brookfield, H. C. and Hart, D. (1971). *Melanesia: A Geographical Interpretation of an Island World.* Methuen, London.

Bulmer, Susan (1964). Prehistoric stone implements from the New Guinea Highlands. *Oceania* **34** (4), 246–268.

Bulmer, Susan (1966). Pig bones from two archaeological sites in the New Guinea Highlands. *Journal of Polynesian Society* **75**, 504–505.

Bulmer, Susan (1973). *Notes on 1972 Excavations at Wãnlek, an Open Settlement Site in the Kaironk Valley, Papua New Guinea.* University of Auckland, Department of Anthropology, Working Paper No. 29. Auckland.

Bulmer, Susan (1975). Settlement and economy in prehistoric Papua New Guinea: A review of the archaeological evidence. *Journal de la Société des Océanistes* **31**, 7–75.

Buschan, Georg (1922–1925). *Illustrierte Völkerkunde* I–III. Globus Verlag G.m.b.H., Berlin.

Chapell, J. (1966). Stone axe factories in the Highlands of east New Guinea. *Proceedings of the Prehistoric Society* **32**, 96–121. London.

Christensen, O. A. (1975a). Hunters and horticulturalists: a preliminary report of the 1972–4 excavations in the Manim valley, Papua New Guinea. *Mankind* **10** (1), 24–36.

Christensen, O. A. (1975b). A tanged blade from the New Guinea Highlands. *Mankind* **10** (1), 37–39.

Clarke, William C. (1971). *Place and People: An Ecology of a New Guinea Community.* Australian National University Press, Canberra.

Cleary, G. J. and Blackburn, C. R. B. (1968). Air pollution in native huts in the Highlands of New Guinea. *Archives of Environmental Health* **17** (5), 785–794.

Conklin, Harold C. (1963). *The Study of Shifting Cultivation: Studies and Monographs.* Unión Panamericana. Washington D.C.

Crosby, Eleanor (1973). A comparative study in Melanesian hafted edge-tools and other percussive cutting implements, Vol. I text, Vol. 2 appendices. Unpublished Ph.D. thesis, Australian National University, Canberra.

Crosby, Eleanor (1977). An archaeologically oriented classification of ethnographic material culture. In *Stone Tools as Cultural Markers: Change, Evolution and Complexity* (R. V. S. Wright, ed.), pp. 83–96. Australian Institute of Aboriginal Studies, Canberra.

Curwen, E. Cecil (1927). Prehistoric agriculture in Britain. *Antiquity* **I**, 261–289.
Curwen, E. Cecil (1930). Prehistoric flint sickles. *Antiquity* **IV**, 179–186.
Curwen, E. Cecil (1935). Agriculture and the flint sickle in Palestine. *Antiquity* **IX**, 62–66.
Damm, Hans (1955). Bericht über die Sonderausstellung "Feldgeräte zur Bodenbearbeitung". *Jahrbuch des Museums für Völkerkunde zu Leipzig* **XIV**, 99–123.
De Schlippe, Pierre (1956). *Shifting Cultivation in Africa: the Zande System of Agriculture*. Routledge and Kegan Paul, London.
Diderot, D. and d'Alembert, J. R. (1751–72). *La Grande Encyclopédie*. Vol. V. Edition exactement conforme à celle de Pellet in quarto. Lausanne et Berne, 1781.
Dosedla, Heinz-Christian (1974). Ethnobotanische Grundlagen der materielle Kultur der Mount Hagen-Stämme im zentralen Hochland von Neuguinea. *Tribus, Veröff. des Linden-Museums*, Stuttgart, No. 23, 155–174.
Dosedla, Heinz-Christian (1975). Steinbeile aus dem Gebiet von Kagua im südlichen Hochlands-Distrikt (SHD) von Papua-Neuguinea. *Tribus, Veröff. des Linden-Museums*, Stuttgart, No. 24, 87–112.
Duff, Roger (1970). *Stone Adzes of Southeast Asia: an Illustrated Typology*. Canterbury Museum, Bulletin No. 3. Christchurch, New Zealand.
Engqvist, H. H. (1944). Læsø Byggeskik. *Nationalmuseets Arbejdsmark* 49–60.
Fenton, Alexander (1978). *The Northern Islands: Orkney and Shetland*. John Donald, Edinburgh.
Freund, Ronald (1968). Agriculture then and now (mimeo). Mukuramanda, PNG.
Gade, D. W. and Rios, R. (1972). Chaquitaclla: the native foot plough and its persistance in central Andean agriculture. *Tools and Tillage* **II** (1), 3–15.
Gardner, Robert and Heider, Karl, G. (1968). *Gardens of War: Life and Death in the New Guinea Stone Age*. Peabody Museum, Cambridge, Mass. and Penguin Books, London (1974).
Glob, P. V. (1951). *Ard and Plough in Prehistoric Scandinavia*. University Press, Århus.
Godelier, Maurice with Garanger, José (1973). Outils de pierre, outils d'acier chez les Baruya de Nouvelle-Guinée: quelques données ethnographiques et quantitatives. *L'Homme* **13**, 187–220.
Golson, Jack (1972). Both sides of the Wallace Line: New Guinea, Australia, Island Melanesia and Asian prehistory. In *Early Chinese Art and its Possible Influence in the Pacific Basin*, Vol. 3, *Oceania and the Americas* (Noel Barnard, ed.). Intercultural Arts Press, New York.
Golson, Jack (1976). The making of the New Guinea Highlands. In *The Melanesian Environment: Changes and Development* (J. H. Winslow, ed.), pp. 45–56. Australian National University Press, Canberra.

Golson, Jack (1977a). Simple tools and complex technology: agriculture and agricultural implements in the New Guinea Highlands. In *Stone Tools as Cultural Markers: Change, Evolution and Complexity* (R. V. S. Wright, ed.), pp. 154–161. Australian Institute of Aboriginal Studies, Canberra.
Golson, Jack (1977b). No room at the top: agricultural intensification in the New Guinea Highlands. In *Sunda and Sahul: Prehistoric Studies in South-East Asia, Melanesia and Australia* (J. Allen, J. Golson and R. Jones, eds), pp. 601–638. Academic Press, London and New York.
Golson, Jack and Hughes, P. J. (1976). The appearance of plant and animal domestication in New Guinea. In *La Préhistoire Océanienne* (José Garanger, ed.), pp. 88–100. Colloque XXII, IXe Congrès de l'Union Internationale des Sciences Préhistoriques et Protohistoriques, Nice, September 1976.
Gorecki, P. P. (1978). Further notes on prehistoric wooden spades from the New Guinea Highlands. *Tools and Tillage* **III** (3), 185–190.
Hammarstedt, N. E. (1914–1915). *Schwedische Opfersteine (Älvkvarnar)*. Beiträge zur Religionswissenschaft II (1–5), herausg. von der Religionswissenschaftlichen Gesellschaft in Stockholm.
Heider, Karl G. (1970). *The Dugum Dani: a Papuan culture in the Highlands of West New Guinea*. Viking Fund Publication in Anthropology No. 49. New York.
Hennig, Egon (1961). Untersuchungen über de Verwendungszweck urgeschichtlicher Schuhleistenkeile. *Alt-Thüringen, Jahresschrift des Museums für Ur- und Frühgeschichte Thüringens* 5, 189–222.
Hose, Charles (1926). *Natural Man: a Record from Borneo*. A. M. S. Press, London.
Hughes, Ian (1977). *New Guinea Stone Age Trade. Terra Australis* 3. Department of Prehistory, Research School of Pacific Studies, The Australian National University, Canberra.
Hvass, Steen (1975). Hodde – et 2000-årigt landsbysamfund i Vestjylland. *Nationalmuseets Arbejdsmark*, 75–85.
Iversen, Johannes (1941). *Landnam i Danmarks Stenalder*. Danm. geolog. Undersøgelser, II Rk., No. 66. C. A. Reitzel, Copenhagen.
Iversen, Johannes (1956). Forest clearance in the Stone Age. *Scientific American* **194**, No. 936, 36–41.
Jarman, H. N. and Bay-Petersen, J. L. (1976). Agriculture in prehistoric Europe – the lowlands. In *The Early History of Agriculture* (J. Hutchinson *et al.*, eds). *Phil. Trans. R. Soc. Ser. B. Biol. Sci*, **275**, No. 939, pp. 175–186.
Jørgensen, Svend (1953). Skovrydning med flintøkse. *Nationalmuseets Arbejdsmark*, 36–43.
Jørgensen, Svend (1974). *Draved* (Report in Danish, mimeo), Nationalmuseets VIII afd. J. No. A 4088. Copenhagen.
Kamminga, Johan (1977). A functional study of use-polished eloueras. In *Stone Tools as Cultural Markers: Change, Evolution and Complexity*

(R. V. S. Wright, ed.), pp. 205–212. Australian Institute of Aboriginal Studies, Canberra.

Kaufmann, Christian (1975). *Papua Niugini. Ein Inselstaat im Werden.* Führer durch das Museum für Völkerkunde Basel, Sonderaustelling.

Koch-Grünberg, Th. (1910). *Zwei Jahren unter den Indianern*, Vol. II. Berlin.

K(rook) J. (1765). *Tankar om swedjande och huruwida det tål inskränkinig.* Stockholm.

Lampert, R. J. (1967). Horticulture in the New Guinea Highlands – C 14 dating. *Antiquity* **41**, 307–309.

Lampert, R. J. (1972). *Hagen Axes: a Pilot Study of Axe Typology in the Central Highlands of New Guinea.* (mimeo). Australian National University, Research School of Pacific Studies, Department of Prehistory, Canberra.

Lampert, R. J. (1977). Kangaroo Island and the antiquity of Australians. In *Stone Tools as Cultural Markers: Change, Evolution and Complexity* (R. V. S. Wright, ed.), pp. 213–218. Australian Institute of Aboriginal Studies, Canberra.

Lea, David M. and Ward, R. Gerard (eds) (1970). *An Atlas of Papua New Guinea.* Collins and Longman, Glasgow.

Lerche, Grith (1969). Kogegruber i New Guineas højland. *Kuml*, pp. 195–209 (English summary).

Lerche, Grith (1977). Double paddle-spades in prehistoric context in Denmark. *Tools and Tillage* **III** (2), 111–124.

Lerche, Grith and Steensberg, Axel (1973). Observations on spade-cultivation in New Guinea Highlands. *Tools and Tillage* **II** (2), 87–104.

Le Roux, C. C. F. M. (1948–1950). *De bergpapoeas van Nieuw-Guinea en hun woongebied*, Vol. I–III. E. J. Brill, Leiden.

Lindt, J. W. (1887). *Picturesque New Guinea.* London.

Master, Kenneth M. (1974). Air pollution in New Guinea: cause of chronic pulmonary disease among Stone-Age natives in the Highlands. *Journal of the American Medical Association* **225** (13), 1653–1655.

Morren, George E. B. (1977). From hunting to herding: pigs and the control of energy in montane New Guinea. In *Subsistance and Survival: Rural Ecology in the Pacific* (T. Bayliss-Smith and R. Feachem, eds). Academic Press, London and New York.

Mountford, C. P. (1941). An unrecorded method of manufacturing wooden implements by simple stone tools. *Transactions and Proceedings of the Royal Society of South Australia* **65**, 312–316.

Müller-Beck, Hans-Jürgen (1965). *Seeberg Burgäschisee-Süd.* Theil 5, *Holzgeräte und Holzbearbeitung.* Mit zwei Beiträgen von F. Schweingruber. Acta Bernensia, Vol. II. Bern.

Müller, P. E. (1887). *Studien über die natürlichen Humusformen und deren Einwirkung auf Vegetation und Boden.* Berlin.

Müller, P. E. (1889). *Recherches sur les formes naturelles de l'humus et leur influence sur la végétation et le sol.* Nancy.

Müller, Sophus (1904). Vej og Bygd i Sten og Bronzealderen. *Aarbøger for nordisk Oldkyndighed og Historie*, pp. 1–64.
Neuhauss, R. (1911). *Deutsch Neu Guinea*, Vols I–III, Berlin.
Nilles, J. (1942–1945). Digging-sticks, spades, hoes, axes and adzes of the Kuman people in the Bismarck Mountains of east-central New Guinea. *Anthropos* **37–40**, 205–212.
Nye, P. H. and Greenland, D. J. (1960). *The Soil Under Shifting Cultivation*. Technical Comm. No. 51, Commonwealth Bureau of Soils. Commonwealth Agricultural Bureau, Farnham, Bucks., England.
Oakley, Kenneth P. (1952). *Man the Tool-Maker*. British Museum, Natural History Books, London.
Petersen, E. Brinch (1970). Ølby Lyng. *Årbøger for nordisk Oldkyndighed*, pp. 5–42 (with English summary).
Pospisil, Leopold (1963). *Kapauku Papuan Economy*. Yale University Publications in Anthropology, No. 76. Holt, Rinehart and Winston, New York, Chicago, San Francisco, Toronto and London.
Powell, Jocelyn (1970). The history of agriculture in the New Guinea Highlands. *Search* (the journal of the Australian and New Zealand Association for the Advancement of Science) **1**, 199–200.
Powell, Jocelyn (1974). A note on wooden gardening implements of the Mt Hagen region, New Guinea. *Records of the Papua New Guinea Museum* **4**, 21–28.
Powell, Jocelyn, Kulunga, A., Moge, R., Pono, C., Zimike, F. and Golson, J. (1975). *Agricultural Traditions of the Mount Hagen Area*. University of Papua New Guinea, Department of Geography, Occasional Paper No. 12. Port Moresby.
Raum, O. F. (1977). The culture historical significance of the Xhosa spade. *Tools and Tillage* **III** (2), 99–110.
Rappaport, Roy A. (1975). *Pigs for the Ancestors: Ritual in the Ecology of a New Guinea People* (7th edn). Yale University Press, New Haven and London.
Robertson, D. S. (1945). *A Handbook of Greek and Roman Architecture* (2nd edn), Cambridge University Press, Cambridge.
Romell, L. G. (1957). Man-made "Nature" of northern Lands. *Proceedings and Papers*. International Union for the Conservation of Nature, 6th Technical Meeting. London.
Romell, L. G. (1967). Die Reutbetriebe und ihr Geheimnis. *Studium Generale* **20** (6), 362–369.
Ross, W. A. (1936). Ethnological notes on Mt Hagen tribes. *Anthropos* **31**, 341–363.
Roussell, Aa. (1934). *Norse Building Customs in the Scottish Isles*. Lerin and Munksgaard, Copenhagen and Williams and Norgate, London.
Salisbury, R. F. (1962). *From Stone to Steel: Economic Consequences of a Technological Change in New Guinea*. Cambridge University Press, Cambridge.
Schier, Bruno (1966). *Hauslandschaften und Kulturbewegungen im östlichen*

Mitteleuropa. 2 Ausg. Otto Schwarz and Co., Göttingen.
Schmidt, Max (1905). *Indianerstudien in Zentralbrasilien*. Berlin.
Schultze Jena, L. (1914). *Forschungen im Innern der Insel Neuguinea*. Bericht des Führers über die wissenschaftlichen Ergebnisse der deutschen Grenzexpedition in das westliche Kaiser-Wilhelmsland 1910. Berlin.
Schwabedissen, H. (1957–1958). Die Ausgrabungen im Satruper Moor. *Offa* **16**, 5–28.
Semenov (= Semyonov), S. A. (1964). *Prehistoric Technology*. Cory, Adams and Mackay, London.
Semenov (= Semyonov), S. A. (1975). 1975-Model Stone Axes. In *Soviet-Union* **7** (304), 39–41.
Steensberg, Axel (1937). Hardsyssels Oldtidsminder. *Hjemstavnsbogen for Ringkøbing Amt*, pp. 24–42.
Steensberg, Axel (1943). *Ancient Harvesting Implements: A Study in Archaeology and Human Geography*. Nationalmuseets Skrifter, Arkæologisk-historiske Række, Vol. I. Gyldendals Boghandel, Nordisk Forlag, Copenhagen.
Steensberg, Axel (1950). De første Bønder og de første Landsbyer. In *Danmarks Historie* (Hvidtfeldt, Koch-Olsen and A. Steensberg, eds), Vol. I, pp. 20–64. Det danske Forlag, Copenhagen.
Steensberg, Axel (1955). Med bragende flammer. *Kuml*, pp. 65–130 (with English summary).
Steensberg, Axel (1957). Some recent Danish experiments in Neolithic agriculture. *The Agricultural History Review*, **V** (II), 66–73.
Steensberg, Axel (1964). A bronze age ard type from Hama in Syria intended for rope traction. *Berytus*, Vol. XV, pp. 111–139.
Steensberg, Axel (1971). Tools and man. In *Man and his Habitat* (R. H. Buchanan *et al.*, eds), pp. 62–78. Routledge and Kegan Paul, London.
Steensberg, Axel (1973a). *Oldtidslevn i New Guineas hverdag*. Forum, Copenhagen.
Steensberg, Axel (1973b). A 6000 year old ploughing implement from Satrup Moor. *Tools and Tillage* **II** (2), 105–118.
Steensberg, Axel (1974). *Den danske bondegård* (2nd edn). Forum, Copenhagen.
Steensberg, Axel (1975). Oldtiden i Nutiden. *Skalk* No. 4, pp. 16–17.
Steensberg, Axel (1976). The husbandry of food production. In *The Early History of Agriculture* (J. Hutchinson *et al.*, eds) *Phil. Trans. R. Soc. Ser. B, Biol. Sci*, **275**, No. 936, pp. 43–54.
Steensberg, Axel (1976). *The Common Fields of Culture* (Grith Lerche, ed.), National Museum, Copenhagen.
Steensberg, Axel (1977a). *Stone Shares of Ploughing Implements from the Bronze Age of Syria: A Contribution to the Early History of the Ard-Plough*. Det kgl. danske videnskabernes selskab, Hist. filosof. Medd. **47** (6). Munksgaard, Copenhagen.
Steensberg, Axel (1977b). Sula, an ancient term for the wheel plough in northern Europe? *Tools and Tillage* **III** (2), 91–98.

Steensberg, Axel (1979). *Draved: an Experiment in Stone Age Agriculture: Burning, Sowing and Harvesting*. National Museum, Copenhagen.

Steensberg, Axel and Østergaard Christensen, J. L. (1974). *Store Valby*, Vol. I–III (with contributions by Tove Hatting and David Liversage in English and an English summary). Det kgl. danske videnskabernes selskab, Hist.-fil. skrifter, Vol. 8 (1), Copenhagen.

Strathern, Andrew and Marilyn (1971). *Self-Decoration in Mount Hagen*. Gerald Duckworth, London.

Strathern, Marilyn (1965). Axe types and quarries: a note on the classification of stone axe blades from the Hagen area, New Guinea. *Journal of the Polynesian Society* **74**, 182–191.

Strathern, Marilyn (1969). Stone axes and flake tools: evaluations from two New Guinea societies. *Proceedings of the Prehistoric Society* **35**, 311–329.

Strathern, Marilyn (1972). *Women in Between: Female Roles in a Male World: Mount Hagen, New Guinea*. Academic Press, London and New York.

Takács, Lajos (1977). A "vész" mint irtási szakkifejezés. *Ethnografia* **LXXXVIII** (2–3), 247–258 (with German and Russian summaries).

Thrane, Henrik (1964). Celtskafter af træ fra Danmarks oldtid. *Årbøger for nordisk Oldkyndighed*, pp. 72–81 (with English summary).

Townsend, William H. (1969). Stone and steel tool use in a New Guinea society. *Ethnology* **8**, 199–205.

Troels-Smith, J. (1937). Beile aus dem Mesolithicum Dänemarks. Ein Einteilungsversuch. *Acta Archaeologica*, **VIII**, 278–294.

Troels-Smith, J. (1953). Ertebøllekultur-Bondekultur. *Årbøger for nordisk Oldkyndighed*, pp. 5–62 (with English summary).

Troels-Smith, J. (1955). Pollenanalytische Untersuchungen zu einige schweizerischen Pfahlbauproblemen, Dissertation. Copenhagen. (Also in *Das Pfahlbauproblem* (W. U. Guyan, ed.). Schaffhausen and Basle 1954).

Troels-Smith, J. (1959). En Elmetræs-bue fra Aamosen og andre Træsager fra tidlig neolitisk tid. *Årbøger for nordisk Oldkyndighed*, pp. 91–145 (with English summary).

Vicedom, G. F. and Tischner, H. (1943). *Die Mbowamb: Die Kultur der Hagenberg-stamme in östlichen Zentral-Neuguinea*. Monographien zur Völkerkunde, 3 Vols. No. 1. Cram de Gruyter, Hamburg.

Vilkuna, Kustaa (1953). Varpå beror den finska svedjebondens kolonisationsförmåga. *Värmland förr och nu*, pp. 9–20.

Vogt, E. (1954). Pfahlbaustudien. In *Das Pfahlbauproblem* (W. U. Guyan, ed.), pp. 119–219. Schaffhausen and Basle.

Waddell, Eric (1972). *The Mound Builders: Agricultural Practices, Environment, and Society in the Central Highlands of New Guinea*. The American Ethnological Society, Monograph 53. University of Washington Press, Seattle and London.

Waddell, Eric (1975). How the Enga cope with frost: responses to climatic perturbations in the Central Highlands of New Guinea. *Human Ecology* **3**, 249–273.

Watson, James B. (1967). Horticultural traditions of the eastern New Guinea Highlands. *Oceania* **38** (2), 81–98.

Weimarck, G. (1953). *Studier över landskapets förändring inom Lönsboda.* Lunds Universitets årsskrift, N.-F.avd., Bd. 48. C. W. K. Gleerup, Lund.

Wheeler, Mortimer (1959). *Early Indian and Pakistan to Asoka. Ancient Peoples and Places.* Thames and Hudson, London.

White, J. P. (1967). Ethno-archaeology in New Guinea: two examples. *Mankind* **6** (9), 409–414.

White, J. P. (1977). Crude, colourless and unenterprising? Prehistorians and their views on the Stone Age of Sunda and Sahul. In *Sunda and Sahul: Prehistoric Studies in South-East Asia, Melanesia and Australia* (J. Allen, J. Golson and R. Jones, eds), pp. 13–30. Academic Press, London and New York.

White, J. P. and Modjeska, Nicholas (1978). Acquirers, users, finders, losers: the use axe blades make of the Duna. In *Trade and Exchange in Oceania and Australia.* (Jim Specht and J. Peter White, eds); *Mankind* **11** (3), pp. 276–287.

White, J. P. and Thomas, D. H. (1972). What mean these stones? Ethno-taxonomic models of archaeological interpretations in the New Guinea Highlands. In *Models in Archaeology* (David L. Clarke, ed.) Methuen, London.

White, J. P., Crook, K. A. and Ruxton, B. P. (1970). Kosipe: a late Pleistocene site in the Papuan highlands. *Proceedings of the Prehistoric Society* **36**, 152–170.

White, J. P., Modjeska, Nicholas and Hipuya, Irari (1977). Group definitions and mental templates. In *Stone Tools as Cultural Markers: Change, Evolution and Complexity* (R. V. S. Wright, ed.), Australian Institute of Aboriginal Studies, Canberra.

Wirz, P. (1924). *Anthropologische und ethnologische Ergebnisse der Central Neu-Guinea Expedition 1921–22.* Nova Guinea, Vol. XVI (1).

Wirz, P. (1952). Die Enga: Ein Beitrag zur Ethnographie eines Stammes im nordöstlichen zentralen Neu-Guinea. *Zeitschrift für Ethnologie.*

Wright, R. V. S. (ed.) (1977). *Stone Tools as Cultural Markers: Change, Evolution and Complexity.* Australian Institute of Aboriginal Studies, Canberra.

Yen, D. E. (1974). *The Sweet Potato and Oceania: An Essay in Ethnobotany.* Bernice P. Bishop Museum, Bulletin 236. Honolulu.

Subject Index